SCIENTIFIC INFLUENCES ON EARLY CHILDHOOD EDUCATION

Scientific Influences on Early Childhood Education offers a new framework for examining the diverse scientific perspectives that shape early childhood education.

As the field takes on an increasing role in addressing children's educational, developmental, and environmental needs, it is critical to more fully understand and appreciate the diverse scientific roots of contemporary early childhood education. This edited collection brings together leading researchers to explain and unpack perspectives that are not often associated with early childhood education, yet have made significant contributions to its development and evolution.

Essential reading for anyone working with young children, this critical and insightful text illuminates the connections between our social values, science, and research in the field.

Dominic F. Gullo is Professor of Early Childhood Education and Director of The Lisa and John McNichol Early Childhood Education Lab at Drexel University, Philadelphia, PA, USA.

M. Elizabeth Graue is Sorenson Professor of Curriculum and Instruction and Director of the Center for Research on Early Childhood Education at the University of Wisconsin-Madison, USA.

SCIENTIFIC INFLUENCES ON EARLY CHILDHOOD EDUCATION

From Diverse Perspectives to Common Practices

Edited by Dominic F. Gullo and M. Elizabeth Graue

NEW YORK AND LONDON

First published 2020
by Routledge
52 Vanderbilt Avenue, New York, NY 10017

and by Routledge
2 Park Square, Milton Park, Abingdon, Oxon OX14 4RN

Routledge is an imprint of the Taylor & Francis Group, an informa business

Library of Congress Cataloging-in-Publication Data
A catalog record for this title has been requested

ISBN: 978-1-138-60516-9 (hbk)
ISBN: 978-1-138-60517-6 (pbk)
ISBN: 978-0-429-46828-5 (ebk)

Typeset in Bembo
by Taylor & Francis Books

CONTENTS

ILLUSTRATIONS

Figures

Tables

CONTRIBUTORS

W. Steven Barnett is a Board of Governors Professor and Senior Co-Director of the National Institute for Early Education Research at Rutgers University. His professional career has focused on the economics of early child care and education including cost–benefits analyses, the long-term effects of preschool programs on children's learning and development, and the distribution of educational opportunities. Dr. Barnett is best known for benefit–cost analyses of the Perry Preschool and Abecedarian programs, research on preschool program design for effectiveness, state pre-K program evaluations, and early childhood policy analysis.

Raquel Bernal is a professor in the Economics Department at Universidad de los Andes. She holds a PhD degree in economics from New York University. Her research focuses on social policy, education, child care, human capital, household decisions, and labor economics. Her recent research investigates the determinants of early human capital accumulation. She has conducted a variety of impact evaluations of early childhood development programs including studies of the short- and long-term effects of both family-home and center-based programs for young children birth to age 5 in Colombia. She is an expert in impact evaluation methodology.

Elena Bodrova, PhD is the Knowledge Advisor at Tools of the Mind. In collaboration with Dr. Leong, she developed the *Tools of the Mind* program based on Vygotskian and post-Vygotskian theories of learning and development. Dr. Bodrova's work on applying Lev Vygotsky's theory to education started in Russia where she worked in the Institute for Preschool Education and continued in the United States where she worked as a visiting professor at Metropolitan State University of Denver and later as a Principal Researcher at Mid-continent Research for Education and Learning. Dr. Bodrova is the author of multiple articles and book chapters on early literacy, assessment, play, and self-regulation. Dr. Bodrova holds a PhD in Child Development and Educational psychology from the Russian Academy of Pedagogical Sciences and an MA in Child Development and Educational psychology from Moscow State University, Russia.

Megan E. Cox has 14 years' experience leading research and technical assistance projects designed to support educators' development and implementation of data-driven continuous improvement and the application of results to improve all children's outcomes. Dr. Cox has expertise in cross-sector early childhood program quality and professional development, assessment design, system and program evaluation, and formative assessment practices. She specializes in using data for continuous improvement in classroom practice and child assessment.

Rebecca Distefano is a PhD Candidate in Developmental Psychology at the Institute of Child Development, University of Minnesota. Her research focuses on the social influences of executive function development in preschool-aged children, with an emphasis on how parents bolster the development of executive function skills in high-risk contexts. Much of her work is done collaboratively with local community partners serving disadvantaged families to inform the design and implementation of interventions to support resilience in children and parents.

Fabienne Doucet, PhD is a Program Officer at the William T. Grant Foundation and an Associate Professor of Early Childhood and Urban Education in the department of Teaching and Learning at New York University (on leave). A critical ethnographer, Doucet studies how taken-for-granted beliefs, practices, and values in the U.S. educational system position linguistically, culturally, and socioeconomically diverse children and families at a disadvantage, and seeks active solutions for meeting their educational needs. Doucet has a PhD in Human Development and Family Studies from UNC-Greensboro and was a postdoctoral fellow at the Harvard Graduate School of Education with fellowships from the National Science Foundation and the National Academy of Education/Spencer Foundation.

David Elkind is Professor emeritus of Child Development at Tufts University in Medford, Massachusetts. He was formerly Professor of Psychology, Psychiatry and Education at the University of Rochester. Professor Elkind obtained his doctorate at UCLA and then spent a year as David Rapaport's research assistant at the Austen Riggs Center in Stockbridge, Massachusetts. In 1964–65 he was a National Science Foundation Senior Postdoctoral Fellow at Piaget's Institut d'Epistemologie Genetique in Geneva. His research has been in the areas of perceptual, cognitive, and social development where he has attempted to build upon the research and theory of Jean Piaget.

Norman Gabriel has recently been appointed as Senior Lecturer in Childhood Studies in the Department of Psychosocial and Psychoanalytic Studies at the University of Essex. He has written one of the first books in Early Childhood Studies, *The Sociology of Early Childhood – Critical Perspectives* (Sage, 2017), to examine young children's lives from a relational sociological perspective. Inspired by the work of Norbert Elias, he is developing a distinctive, multi-disciplinary approach to childhood that is historical and comparative. He is currently researching how psychoanalytic theories can be enriched by relational sociology to understand the socio-emotional development of young children in different societies.

Ellen Galinsky is the chief science officer and Executive Director of Mind in the Making at the Bezos Family Foundation. She also serves as President of Families and Work Institute. Over her career, she has conducted research on child care, parent–professional relationships, parental development, and work–family issues. She is the author of the best-selling *Mind in the Making: The Seven Essential Life Skills Every Child Needs*. Other highlights include writing more than 100 books/reports and over 300 articles, serving as a child care expert on Dr. T. Berry Brazelton's television series and as a parent expert in the *Mister Rogers Talks to Parents* TV series; and being the elected President of the National Association for the Education of Young Children. In 2018, the Work and Family Researchers Network announced that the Ellen Galinsky Generative Researcher Award will be given on an ongoing basis.

Sally Campbell Galman is professor of Child and Family Studies at the University of Massachusetts-Amherst College of Education. As an anthropologist of childhood, her research focuses on the study of children and gender with a focus on the school experiences of young children who are transgender or otherwise gender diverse. She is an Editor-in-Chief of *Anthropology and Education Quarterly*, a section editor of the journal *Jeunesse: Young People, Texts, Cultures*, and is the author of the award-winning *Shane* series of qualitative methods texts.

M. Elizabeth Graue is Sorenson Professor of Curriculum & Instruction and Director of the Center for Research on Early Childhood Education at the University of Wisconsin – Madison. A former kindergarten teacher, she studies early childhood policy and practice, with particular attention on the many forces that shape teaching. This focus has motivated studies on community beliefs about readiness, home school relations, class size reduction, professional development on early math pedagogy, and enactment of public prekindergarten policy.

Dominic Gullo is a professor of early childhood development and education at Drexel University in Philadelphia. He has worked in the profession for over 35 years as a Head Start and public-school classroom teacher as well as a university professor and researcher. Dom's areas of specialization and research include early childhood curriculum and assessment, school readiness, risk and resiliency in early childhood, and early childhood school reform. Dom has held many leadership positions in the field with professional associations and has consulted widely with schools throughout the United States and abroad. He has published widely and presented his work both nationally and internationally.

Kathleen Hebbeler is a Senior Principal Education Researcher with SRI International. She has been conducting research and evaluation of education, health, and social programs for young children and their families for over 30 years. Her most recent work involves supporting state agencies in using data for program improvement. Her areas of expertise include home visiting, child development, early intervention, general and special education, assessment, standards, accountability, and community collaboration. As a member of the National Research Council's Committee on Developmental Outcomes and Assessment for Young Children, she was a contributing author to the 2008 report, *Early Childhood Assessment: Why, What and How?*

Magdalena Janus has a PhD from Cambridge University, UK and is a Professor at the Department of Psychiatry and Behavioural Neurosciences and Offord Centre for Child Studies at McMaster University. She is a co-author, with the late Dr. Dan Offord, of the Early Development Instrument (EDI), a measure of children's developmental health at school entry. The span of her research includes early development and health of all children, social determinants of children's health, and measurement and indicators of child development. She studies children in Canada as well as globally with organizations such as the World Bank, WHO, UNICEF, and UNESCO.

Jacqueline Jones is the President and CEO of the Foundation for Child Development. During the first term of the Obama administration, Dr. Jones served as Senior Advisor on Early Learning to Secretary of Education Arne Duncan and as Deputy Assistant Secretary for Policy and Early Learning. Prior to federal service she was the Assistant Commissioner for the Division of Early Childhood Education in the New Jersey State Department of Education. For over 15 years Dr. Jones worked as a Senior Research Scientist at Educational Testing Service. She attended Hunter College and earned an MA and a PhD from Northwestern University.

Sharon Lynn Kagan is the Virginia and Leonard Marx Professor of Early Childhood and Family Policy and Co-Director of the National Center for Children and Families at Teachers College, Columbia University, and Professor Adjunct at Yale University's Child Study Center. Recognized internationally and nationally for her accomplishments related to the care and education of young children, Kagan is a prolific pubic speaker, the author of 325 articles and 18 books, and a member of over 30 national boards or panels. The recipient of countless research awards, Kagan focuses on the development of new knowledge that impacts governments and institutions as they craft policies for young children, according particular attention to systems theory, governance, finance, pedagogy, and transitions. Drawing on leadership experiences as a public school administrator and as the Director of the New York City Mayor's Office of Early Childhood Education, as well as from her policy work in 78 countries around the world, Kagan has been instrumental in altering policies affecting young children and families.

Eva Landsberg is a PhD student in History at Yale University. She was previously a research assistant at the National Center for Children and Families (NCCF) at Teachers College, Columbia University, where she was involved in The Early Advantage study of high-performing international early childhood education and care systems. While at NCCF, she co-authored multiple book chapters and articles on systemic issues of quality and equity in ECEC. Prior to joining NCCF, she conducted research as a policy intern at the Children's Defense Fund in Washington, DC, specifically focusing on the intersection of poverty and early education. She received her BA in History from Yale.

Deborah J. Leong, PhD is President and Cofounder of Tools of the Mind (*Tools*) with Dr. Elena Bodrova. The *Tools* intervention nurtures executive functions through the development of intentional mature make-believe play and playful learning. *Tools* has several technology solutions for professional development and helping children learn to read. She has co-authored numerous books, articles, and educational videos on the Vygotskian approach to psychology, the development of play, and early childhood assessment. Deb is professor emerita of Developmental and Cognitive Psychology at Metropolitan State University of Denver. She received her PhD and BA from Stanford University and her MEd from Harvard University.

Jessica Navarro has studied at Brown and Washington University, and is currently a doctoral student in the Human Development and Family Studies Department at the University of North Carolina Greensboro. Inspired by her previous career as a clinical social worker, she is interested in translational research related to moral and character development with both families and schools.

Salmi Noor completed her Bachelors in Health Sciences and a Masters in Health Research Methodology at McMaster University, Hamilton, Canada. For her MSc thesis, Salmi examined the association between school level prevalence of special needs in kindergarten and developmental health outcomes of children based on neighbourhood socioeconomic status. Salmi is currently a year medical student at the University of Ottawa and her research interests include paediatric and women's health outcomes. She hopes to combine her passion in health sciences and medicine with her training in research methods to improve overall health and help mitigate disparities facing vulnerable populations.

Milagros Nores, Co-Director for Research at NIEER, oversees research operations at NIEER, while pursuing her own research. Her expertise and interests are in early childhood development, data-driven policy development, evaluation design, economics of early childhood, and English language learning. Dr. Nores has an early childhood study in Colombia, a study on parental–child educational practices for minority children in the United States, and is a lead in evaluations of the Philadelphia, Seattle, and West Virginia's preschools programs. She has a PhD in Education and Economics from Columbia University and an EdM in Educational Administration and Social Policy from Harvard University.

Sharon Ryan is Professor of Early Childhood Education at the Graduate School of Education and a Research Professor at the National Institute of Early Education Research, at Rutgers University. Dr. Ryan uses a range of mixed methods designs to research early childhood curriculum and policy, and workforce issues including teacher education and professional development. Dr. Ryan has worked in the early childhood field in Australia as a preschool teacher, program leader, curriculum advisor, and special educator.

Ayesha Siddiqua is a PhD candidate in the Department of Health Research Methods, Evidence, and Impact at McMaster University. Her thesis examines the association between socioeconomic status and the health of children with Autism Spectrum Disorder. Ayesha has graduate training in quantitative, qualitative, and mixed method research methodologies. Her research interests include modifiable determinants of health, health services evaluation, big data analytics, health economics, and health policy development.

Jonathan Tudge was educated at the Universities of Lancaster and Oxford (UK) and Cornell (USA), where he studied with Urie Bronfenbrenner. He is a professor of human development at the University of North Carolina at Greensboro. His books include *The Everyday Lives of Young Children: Culture, Class, and Child Rearing in Diverse Countries* (2008) and *Developing Gratitude in Children and Adolescents* (2018, co-edited with Lia Freitas). He has co-edited two other books related to culture and youth development. He has also published widely on Bronfenbrenner's theory – how it developed and how it can best be applied in research.

Phil Zelazo is the Nancy M. and John E. Lindahl Professor at the Institute of Child Development, University of Minnesota. His research has helped shape current understanding of executive function (EF) and its development, including the key roles of reflection, rule use, hierarchical complexity, mindfulness, and emotion (hot versus cool EF), and it has led to the design of widely used standardized measures of EF (e.g., he was lead developer of the EF measures for the NIH Toolbox measures of EF) and the creation of effective interventions for promoting the healthy development of EF in early childhood. Zelazo has received numerous awards, including a Boyd McCandless Young Scientist Award from the APA, and a Canada's Top 40 Under 40 Award. He was Founding Editor of the *Journal of Cognition and Development*, he serves on several editorial boards, and is editor of the two-volume *Oxford Handbook of Developmental Psychology* (2013).

FOREWORD

Barbara Bowman

Since early in the twentieth century, scientific research on child development has increasingly formed the base for early education pedagogy. From the child study movement, to the ethological studies, to neurological scans, science has added immensely to our understanding of children and their later educational achievement and social accomplishments. Theories and research on child development have been critical to understanding the influence of child rearing on later child accomplishments including school achievement, social competencies, and health. However, the sciences also have created a fragmented knowledge base with each science looking at children from its own perspective. This has been a source of confusion and conflict in the broader early childhood field. Questions like, "When should Johnny be excluded from school?" or "Should a 4-year-old learn the alphabet or count to 100?" reveal conflicts between professional organizations, local pediatricians, therapists, and preschool programs, each using science to validate their own perspective. As evidenced by the chapters in this volume, with all good intentions, scientific research and perspectives have often resulted in the fragmentation of the child. The results being that evidenced based practice in early childhood may target specific child attributes and developmental domains, while those not targeted have diminished emphasis.

This volume is designed to review together the various sciences that have influenced our thinking about children and the practices that are associated with their education and care. By reviewing this research in one place, we can gain insights into the evolution and possible integration of the complex of ideas that form the knowledge base for the field. Despite outdated ideas (don't kiss the baby), and mistaken interpretations (Montessori valued free play), science can contribute to our understanding of development and learning and on early care and education. Keeping abreast of the research in each of these sciences is an essential for practicing professionals.

It is easy to think of each of the sciences in its own pigeon-hole, isolated from other aspects of child development or early education and family and community life. Yes, each is inextricably tied to the other and to the larger economic, social, and political systems in which they are embedded. By bringing together the biological, political/economic, social, educational, and psychological sciences, we can view the knowledge base as it functions in the real world of competitive ideas. By understanding the sciences in the context of interconnected micro- and macrosystems that drive modern life, our research should support policies that better meet our goals for children.

Perhaps, most importantly, this volume addresses one of the central issues of our time: the knowledge base that responds to the new societal value, equality. Instead of reflecting the status quo, the early childhood profession is increasingly seen as the spearhead for equity for people of color, low income families, children with special needs, and others from diverse cultural communities. Small model programs and Head Start demonstrated that early education can confer life-long educational advantages to poor children. Equality is now central to the commitment to preschool education as well as other programs associated with early childhood, and research is needed to guide its implementation in a manner that ensures that all children have opportunities to take part in high quality early childhood education and care experiences that will optimally support their development and learning potential.

Traditionally, theories and research have reflected the unconscious biases of the mainstream society. Samples drawn from one socio/cultural community were applied to others, without consideration of the social and political context. Some group practices were privileged because of the power of their adherents; others were ignored for the lack of power. Science often reinforced the notion that the goal of development and education is not just normative, but white middle class normative. Consequently, difference is portrayed as a deficit and intervention as remedial.

Rather than reinforcing traditional perspectives, the authors in this volume report on child development research that is responsive to cultural differences, recognizing there are different ways to raise healthy children. They also apply the cultural lens to illuminate new or more successful ways to mitigate or eliminate resource inequalities. The authors also reflect cultural diversity, giving personal as well as intellectual authenticity to their work.

Using science to help drive education is likely to increase with the advances in technology, artificial intelligence, and assessment strategies. It is also likely to be more contentious as new ideas conflict with earlier ones. It is essential for professionals to develop a sound knowledge base that can be viewed against a backdrop of social values. Science is not value free. Scientific research is not value free. We will need to be intentional and vigilant to ensure that new research reflects the field's commitment to diversity and the connection between values, science, and research is transparent.

1

CHARACTERIZING THE SCIENTIFIC INFLUENCES ON EARLY CHILDHOOD EDUCATION

Dominic F. Gullo

Early childhood education (ECE) is complex in both its theoretical underpinnings and their interpretation into practice. ECE spans diverse disciplinary perspectives that include, but are not limited to, psychology, health/medicine, management, sociology, policy, and anthropology. Each of these disciplines has contributed a unique scientific understanding to ECE that has contributed to its dynamic evolution. Diverse policies that frame services and practices in ECE reflect the intermingling and coalescing of these scientific perspectives.

The patchwork of scientific knowledge woven into the fabric of ECE has to some extent created a liminality within early childhood education, leading some to query whether ECE is a field or a profession. According to Bornfreund and Goffin (2016), ECE is not a profession.

> The early childhood education field does not conform to the standards of organized professions nor is it held accountable as such, as reflected in the variability in teachers' knowledge and skills. Consequently, although central to its aspirations, the early childhood education field cannot yet claim status as a profession. To pretend otherwise is a disservice to families and their children.
>
> (*p. 1*)

A lack of a well-defined professional identity along with the diversity of foundational perspectives that frame ECE are manifest not only in the individuals who comprise the field, but also in the structure of the field itself, which is often characterized as being a fragmented, episodically funded non-system. This characterization is symptomatic of the complexities inherent in the needs reflected in addressing young children's learning, development, and well-being, each of which is unique and requires focused consideration at different junctures in the early childhood developmental spectrum. These complexities have led to distinct, but often separate programs and services, each based on disparate underlying supportive tenets.

The practice and research foci of ECE has been influenced by scholars who represent a multiplicity of scientific, theoretical, and disciplinary perspectives. This is becoming increasingly the case as the scope of ECE widens. It is therefore very important for:

a ECE researchers to understand how different scientific and/or theoretical perspectives have conceptualized ECE within their unique views of the world;

b scholars in other traditions to recognize and understand how their perspectives are embraced (or not embraced);
c ECE practitioners (particularly those in teacher education) to understand how both (a) and (b) have operated.

The consequences of the disintegration in ECE, in conjunction with a lack of clarity regarding definition of purpose, give rise to ECE being characterized as having a range of structural, curricular, pedagogical, and evaluative approaches at all levels and in all contexts – from the immense variations within and across settings that serve young children, to the preparations of ECE professionals, to the beliefs and practices of ECE policy makers and scholars. It is little wonder then that there are diverse scientific and theoretical touchstones and beliefs about research methodology and evidence that result in splintering the early childhood community in ways that make referring to early childhood education something that is fraught with uncertainty. Ultimately, some may wonder if there is a coalescing scientific *linqua franca* of early childhood education. Still others may question if there should be. To begin understanding these tensions, one must first ask the question, "What is science?"

What is Science?

The question, "What is science?" has been posed for centuries. The answers to the question are diverse and speak to wide-ranging scientific fundamentals. These fundamentals address the characteristics of scientific thought, practices surrounding scientific thought, and how scientific thought can be used in different ways to accomplish different goals. According to Clough and Kruse (2009), "Well-established science knowledge is durable, but always open to revision" (p. 3). Well-established scientific knowledge is often thought of as proven truth, but, in actuality, one can never know if scientific knowledge is absolute truth. "Truth" depends on the temporal and contextual attributes under which it is established.

Across various scientific disciplines, there are the basic or pure sciences and the applied sciences. Basic sciences are those that produce rudimentary but essential information that could be used to predict and/or explain phenomena that exist in the natural world. The applied sciences, on the other hand, are those that use the information generated by the basic sciences in ways that spawn practical applications of that information and/or knowledge. If a science of early childhood education exists, it could be considered in that category of sciences. The distinction between applied science and other forms of science has been the focus of discussion for many decades.

According to Feibleman (1961), science has two purposes, "to know" and "to do." Basic science or pure science is used as a method to investigate nature to satisfy the need to know (understanding). Applied science is the use of the knowledge generated by basic science for some practical human purpose (action).

Education, including early childhood education, can be characterized as an applied science. Within educational settings, the authoritative intent of applied science is to shape the approach that professionals use to influence the thinking and capabilities of students (Cillers & Timmermans, 2014). The focus of the scientific perspectives that inform educational routines has necessarily shifted to practice-relevant research. According to Ulrich (2008), "the practical implementation (of knowledge) required more than scientific training, as a thorough understanding of the concrete situation to which such methods are applied is needed."

With yet another perspective, Aumann (2011), a Nobel Laureate in economic sciences, stated in his Nobel Laureate Symposium that

> there is no such thing as "pure" or "applied" science ... there is such a thing as science, but there is no difference between pure and applied science. Science is one entity and cannot be separated into different categories .... There is no such thing as exclusively "pure" or "applied" science, only *good* science.
>
> *(pp. 1, 3)*

He concludes by saying that one must follow the path of one's inquisitiveness and follow the principles that regulate scientific inquiry.

According to Understanding Science (2019), there are a number of key characteristics that exemplify science and scientific inquiry. These characteristics can be used as criteria to determine how scientific a body of knowledge may be. They include:

- Science focuses on the natural world (as opposed to the supernatural world).
- Science uses testable ideas.
- Science relies on evidence.
- Science involves the scientific community.
- Science leads to ongoing research.
- Sciences benefits from scientific behavior.

In the western world, it was during the sixteenth century that the word *science* came into commonplace usage (Sample, 2009). While most people have a sense of what science is, a single definition is not easy to come by. In fact, in 2009, the Science Council of Great Britain spent a year developing a definition of science. According to the Science Council, "Science is the pursuit of knowledge and understanding of the natural and social world following a systematic methodology based on evidence" (Science Council, 2009). Systematic scientific methodology includes:

- objective observation through measurement and data collection (not necessarily using mathematics as a tool);
- evidence;
- experimentation and/or observation as standards for hypothesis testing;
- inductive reasoning to determine generalizable rules and/or conclusions;
- replication to determine validity of findings; and
- critical exposure to scrutiny.

The question that must be asked is to what extent can early childhood education be characterized by the scientific knowledge and thought that influenced and formed its structure and practice; and does this characterization fit the criteria that determine authoritative scientific quality? In the following section, a brief overview of the history of scientific thought that influenced, shaped, and molded early childhood education will be examined. As will become evident, the dynamic elements of what early childhood education was, and is, are influenced by a diverse set of scientific thinkers.

History of Early Childhood Scientific Thought

Early childhood education has a long history of borrowing from other fields as it constructed the tenets, systems, and principles around which it evolved and operated. Concomitantly, thinkers from other fields have freely used the tools of their discipline to weigh in on the contexts and strategies used to support young children's learning, development, and well-being; this input has shaped the collective philosophy and practices that today we refer to as early childhood education. Each of these historic and notable influences made significant contributions to the prevailing period-specific early childhood theory and practices (Morrison, 2018). Individuals who contributed to these influential viewpoints represented both the social sciences (e.g., theology, philosophy, psychology) and the biological sciences (e.g. medicine, neuroscience). In addition, the theoretical underpinnings that came to be associated with ECE emanated from these scientific influences as well. Reflected across ECE's scientific as well as theoretical histories are three distinct, albeit interrelated viewpoints (Woodhead, 2006), each of which embodies multidisciplinary scientific and theoretical ideas.

The first is a *developmental perspective*. It stresses the significance of young children's physical and psychological growth and maturation. This perspective emphasizes the regularities and universalities in children's development, as well as their dependencies and vulnerabilities. The genesis and nucleus of this perspective can be traced back to at least as far as Plato (428–348 BC), who asserted,

> And the first step ... is always what matters most, particularly when we are dealing with the young and tender. This is the time when they are taking shape and when any impression we choose to make leaves a permanent mark.
>
> *(cited in Clarke & Clarke, 2000, p. 11)*

Emblematic of this perspective, philosopher and theologian John Amos Comenius (1592–1670) posited that children have a timetable for development and that knowledge should be presented to them based on readiness. He emphasized that one needs to consider the ways in which children instinctively learn, rather than teaching through memorization of text that children do not understand.

John Locke (1632–1704), a physician and philosopher, advanced the idea that the quality of the environment will impact the quality of children's learning and development. While most famously associated with the term "tabula rasa," his views on children's education became quite progressive.

> Children are strangers to all we are acquainted with. They must play. Their minds wander. They need to be busy, and they love change and variety. They are naturally curious. To motivate, the skillful teacher simplifies lessons, sympathetically answers naïve questions, seizes the moment when the child is in tune, engaged, and responsive.
>
> *(cited in Gibbon 2015, p. 23)*

Maria Montessori (1870–1952), an Italian physician, developed a child-centered approach to education that is based on the belief that children are eager for knowledge and capable of initiating learning in a supportive and thoughtfully prepared environment. Her work in early childhood education was based on scientific method.

Swiss clinical psychologist, Jean Piaget (1896–1980) strongly influenced early education and development theory and practice through his pioneering work in genetic epistemology. Piaget asserted that knowledge is constructed through direct and active interaction with the environment. He emphasized that both nature (one's biological dispositions) and nurture (one's physical and social environments) influence development and learning in children.

Regardless of the specified perspective, the following themes are consistently represented in the psychological science's developmental perspective (Woodhead, 2006):

- Young children's developmental functioning in all domains is characteristically different from that of older children.
- Several developmental transitions occur in young children's functional capacities that begin in infancy and proceed through the years of early schooling.
- Early childhood is the time when children are the most dependent on others for both their survival and for emotional security, social assimilation, and intellectual and cultural proficiencies.
- Children's development in the early years is particularly vulnerable to negative influences from their physical and social environments. The impact of these negative influences can be suffered throughout childhood and into their adult years.
- While children's development can be described in terms of wide-ranging universal tenets, individual differences and variability in development can occur due to mutable influences from their physical and social environs.

The second is a *political and economic perspective*. This perspective, while enlightened by the developmental perspective, transforms scientific knowledge into social and educational interventions and practices. Much of the emphasis of this perspective is testing the hypothesis that intervening during the early years can do much to overcome the negative effects of deleterious influences in children's lives. It is based on the understanding that developmental processes in the early years are time-sensitive. There are periods during which environmental stimuli have greater impact on children's development; leading to greater vulnerability (Shonkoff & Phillips, 2000; Shonkoff & Levitt, 2010). According to Boyden and Dercon (2012),

> Children deserve to be the focus of policy attention, but because they stand to benefit immensely from appropriate and effective intervention and also because investing in children lays the foundation for the future of a country. Without these investments, economic growth cannot be sustained.
>
> *(p. 46)*

Throughout history, a diverse set of scientific thinkers have exemplified this perspective. Johann Pestalozzi (1746–1827), who was an educational reformer and pedagogue, asserted that "formal education" within a school setting is necessary for children to be able to integrate knowledge of home life, vocational education, and literacy. The goal of education for Pestalozzi was to educate the "whole child" by balancing educational experiences between heart, hands, and head. In this way, children can and should learn through hands-on activities with concrete objects.

Social reformer and inventor Robert Owen's (1771–1858) beliefs and practices shaped his conviction that early education is in the best interests of society. He created the basis for the infant school movement. Owen encouraged and promoted children's engagement in free unstructured play, supported the idea of learning from concrete objects and nature, and advocated for education outside the home.

Around the same time, Friedrich Froebel (1782–1852), a pedagogue who laid the foundation for modern education, contributed to the idea that formal curriculum, instructional methods, and teacher training are necessary for supporting children's learning, emphasizing that this was particularly essential during the early years. He is credited with pioneering the concept of kindergarten.

Jane Addams (1860–1935) was a political activist, social worker, and co-founder of the ACLU. She worked to reduce child labor in order to improve educational opportunities for young children. She is recognized as the founder of the social work profession in the United States. She is well known for saying that "America's future will be determined by the home and the school. The child becomes largely what he is taught; hence we must watch what we teach, and how we live" (Jane Addams quotes, n.d.).

Illustratively, many of the scientific findings represented in this perspective are what ultimately led to the 1960s' early childhood intervention efforts resulting from the "War on Poverty."

> The assumption at the time was that the results of scientific research and evaluation would ultimately be so precise as to allow social scientists to determine which programs and policies were worthy of investment of public funds and citizen energies. Proven models would be described, disseminated and ultimately cloned.
>
> *(Schorr, 2004, p. xvii)*

The third is a *social and cultural perspective*. Within this perspective, early childhood education and development are viewed within the context of society and culture. Early childhood education and development cannot be understood unless the society and culture in which they evolve is understood. This perspective arose out of the critique of the strictly and narrowly interpreted developmental paradigm that shaped both practice and policy and became a dominant force in ECE.

To illustrate this point, "Developmentally Appropriate Practice" (DAP: Bredekamp, 1987) as it was initially conceived reflected a strong developmental perspective. In part, it was developed as a "scientific defense" of informal, child-centered, and play-based programs for young children (Woodhead, 2006), reinforced by Piagetian theory. It emphasized:

- deference for universal stages of development;
- activity-based learning;
- exploration; and
- a skilled practitioner that guided and facilitated learning.

In another example, and counterbalancing the mainstream, mostly male, history of early childhood education, is work from a critical or reconceptualist perspective. It was designed to decenter the assumptions that, for many, serve as the field's foundation. This work, which spans the end of the twentieth and beginning of the twenty-first centuries, calls into question a unitary developmental narrative that propelled professional statements like *Developmentally Appropriate Practice* (Bredekamp & Copple, 1997). DAP in its original iteration was eventually criticized as not being sensitive to the cultural diversity reflected in children's family and community experiences (e.g., Mallory & New, 1994). In 1997, a revised DAP was published (Bredekamp & Copple, 1997). In this edition, a revised 12-point position statement was published. It included point number 6, which stated, "Development and learning occur in and are influenced by multiple social and cultural contexts" (p. 12).

Depending on the theorist, critiques focus on the lack of attention payed to the role of culture in development (Lubeck, 1996), the ways that traditional early childhood practices reflect the lives of affluent children, or the role that power plays in education (Ryan & Grieshaber, 2005). These groups that respond to the mainstream discourse shift and change over time to counterbalance the certainty in policy-relevant research and provide new ways of examining our practice. Throughout history, a number of diverse scientific revisionists, centered on the socio-cultural aspects of education, have contributed to what would become common practice in ECE. Representative of these are those discussed below.

Martin Luther (1483–1546), a theologian and social reformer, advocated for communities playing a crucial role in ensuring that all young children, including girls, were educated. Luther believed that teaching literacy was the primary role of education.

In the eighteenth century, the philosopher, writer and composer Jean-Jacques Rousseau (1712–1778) held the idea that children should be nurtured to develop their own strengths in an atmosphere that is absent of interference or restrictions; that children were qualitatively different from adults and developed in stages; each stage having its own unique characteristics and supportive needs. For young children, he advocated for a heuristic approach method of discovery learning.

The McMillan sisters, Margaret (1860–1931) and Rachel (1859–1917), were social reformers and active in British politics (Liebovich, 2016). Born in the United States, they both moved to Scotland upon the death of their father. They spent their lives focused on levelling the playing field for children in poverty who resided in London; first focusing on health and hygiene, and then on education. This resulted in their opening what would become a nursery school to ameliorate the conditions they saw as detrimental to the well-being of the young children whom they served. They become pioneers in the nursery school movement and influenced the same in the United States.

Lev Vygotsky (1896–1934), a Soviet psychologist, was the originator of the theory of human cultural and bio-social development. He saw psychological development as the result of an individual's interpersonal connections and actions within a social milieu.

Finally, Urie Bronfenbrenner (1917–2005), a Russian-born American psychologist is best known for his ecological systems theory of child development. His perspective was pivotal in changing the focus of developmental psychology by drawing attention to the large number of societal/cultural and environmental influences on child development.

The contributions of each of these historic figures with diverse scientific viewpoints on young children and education are still apparent today in early childhood development and practice. Without them, ECE would not have evolved into what it is today.

The Evolution of Early Childhood Education

As the field of early childhood education developed and evolved, it ultimately was recognized as a necessary component of children's lives. Therefore, it seems fitting at this time to provide a definition of early childhood education as it exists today. According to the National Association for the Education of Young Children (NAEYC, 2009), ECE is the education and care of children from birth through age 8. Given NAEYC's definition, ECE professionals work with various age groups of children. These include:

- Infants and toddlers: birth to 36 months.
- Preschoolers: 3- and 4-year-olds.
- Kindergartners: 5- and 6-year-olds.
- Primary grades: 6-, 7-, and 8-year-olds (1st, 2nd, and 3rd grades).

Due to ECE being defined broadly across these age groups, the settings in which ECE professions work is equally diverse, and include:

- center-based and home-based child care;
- preschool settings (i.e., public and private preschools, Headstart);
- public and private kindergarten classrooms; and
- public and private primary grade schools and classrooms.

Early childhood education propagated a myriad of programs and services for children and families. As a result, it is not surprising that we have service fragmentation that marks early childhood education as one of the most disconnected professional fields that exists today. At the same time, program expansion produced a focus on evaluation and research to ensure that public investment was producing results. The field drew scholars from economics, public policy, political science, psychology and sociology, and neuroscience into the early childhood domain. This often times resulted in scientific impositions on ECE that would both shape and change its structure and purpose.

Given these diverse theoretical, disciplinary, and practice roots, it is not surprising that the scientific base that emerged to investigate young children and the practices that support their development has been equally diverse. As this research emerged, it reflected researchers' disciplinary assumptions and produced programmatic and structural components, rendering early childhood education more complex and increasingly fragmented.

While it is certainly the case that ECE is complex enough to warrant different tools and viewpoints, it is clear that the study of young children's education and care experiences has no integrated theory or science that serves as a universally agreed upon conceptual and/or theoretical framework. Instead, ECE research, policy and practice is informed by a kind of kaleidoscope of sciences that embody the idea that what you see depends on how, where, when, and why you look. The pivotal question is: has ECE borrowed from the vast domain of sciences to justify its practices and belief system, or has ECE constructed a new integrated scientific body of knowledge that is distinctive and discernibly set apart from those sciences that contributed to it?

With this historical background in mind, and predicated on the belief that if someone talks about a science of ECE, we need to ask, "Whose science?" It recognizes that early childhood education is multiply determined, and that programs, policies, and research claims have baked into them the different scientific traditions that have contoured the early childhood community and its discordant services. We argue that to understand early childhood education in the twenty-first century, we need to have a better sense of ECE's scientific genealogy.

Developing Scientific Perspectives of Early Childhood Education

In this volume, leading scholars in their respective scientific fields have described how various scientific outlooks have shaped and influenced ECE research, structure, and practice. Each of the authors, to varying degrees, has described:

- The relationship between their science and ECE.
- The significance each science has as it connects to:
    - changing expectations for ECE;
    - changing practice; and
    - changes in research focus;
- The implications of the science for advancing ECE or modifying ECE to adjust to the needs of a special population or environmental shift.

Early education has taken on an increasing role in addressing children's educational, developmental, and environmental needs. Therefore, there is a need to more fully understand and appreciate the diverse scientific pathways contemporary early childhood education has traveled. In order to accomplish this task, three sets of queries should be considered.

First, there is a need to put the research in perspective related to the synergy that exists between ECE and the sciences that influenced it. As such, the following questions should be addressed:

- How have different scientific traditions "conceptualized" or "reconceptualized" early childhood education?
- How is understanding the scientific and intellectual foundations of ECE a critical component in advancing the knowledge generating activities related to ECE?
- How can early childhood education build on its duly recognized foundation of applied developmental psychology in appreciating and supporting the other scientific traditions that have shaped it?

Second, there is a need to examine if and how the research from various scientific fields has contributed to early childhood practice by asking:

- In what ways have the different scientific traditions informed early childhood education theory, philosophy, practices and structures?
- What are the mechanisms that will channel research findings from other disciplines into early childhood? Should we work to create and/or elevate early childhood education into a scientific field with its own identifiably research identity? If so, how?
- How have some disciplines misunderstood the dynamics of ECE and how can ECE support a better understanding of its practice?

Third, there is a need to identify, describe, and define the sciences of early childhood education and the research that distinguishes it. The following questions are posed as a means of response to this issue:

- What role does research from other sciences play in examining this issue?
- What are the distinguishing characteristics of early childhood research that differentiate it from research of other fields or sciences?
- What are the research traditions in early childhood education that contribute to its distinctive identity?

Each of the chapters that follow will offer an extensive scientific framework for examining the diverse perspectives that have shaped early childhood education. Each will examine ECE from the scientific perspectives represented in this volume. They include the psychological sciences, biological sciences, social sciences, system sciences, and educational sciences. Representing the psychological sciences will be chapters describing the impact of developmental perspectives and socio-cultural perspectives. The biological sciences are represented by chapters on medicine and neurosciences and their impact on ECE. The impact of the social sciences on ECE includes chapters on anthropology, sociology, policy, and economics. Systems sciences will include chapters on the impact of bioecological systems and organizational systems. Finally, the educational sciences will be represented by chapters on pedagogical and assessments sciences.

This volume is not meant to definitively answer the question, "Is there a science of early childhood education?" Rather, it is meant to generate thought-provoking discussion related to the questions posed above. Hopefully, this discussion will lead to the conclusion that early childhood education, whether it's a field or a profession, or neither, has a rich scientific genealogy.

References

Aumann, R.J. (2011). Pure and applied science. *Rambam Maimonides Medical Journal*, 2(1): e0017. doi:10.5041/RMMJ.10017

Bornfreund, L., & Goffin, S. (2016, September). Early childhood education is not a profession. *Pacific Standard*. Retrieved from https://psmag.com/news/early-childhood-education-is-not-a-profession

Boyden, J., & Dercon, S. (2012). *Child development and economic development: Lessons and future challenges*. Oxford: Young Lives.

Bredekamp, S. (Ed.) (1987). *Developmentally appropriate practice in early childhood programs serving children from birth through age 8*. Washington, DC, National Association for the Education of Young Children.

Bredekamp, S., & Copple, C. (Eds.) (1997). *Developmentally appropriate practice in early childhood programs*. Washington DC, National Association for the education of Young Children.

Cillers, E.J., & Timmermans, W. (2014, September). Applied science and education: Vice versa benefit. Paper presented at the International Conference on Education and Social Sciences, Singapore.

Clarke, A., & Clarke, A. (2000). *Early experience and the life path*. London, Jessica Kingsley.

Clough, M.P., & Kruse, J.W. (2009). Characteristics of science: Understanding scientists and their work. The story behind the science. Retrieved from https://www.storybehindthescience.org/pdf/chara cteristics.pdf

Feibleman, J.K. (1961). Pure science, applied science, technology, engineering: An attempt at definitions. *Technology and Culture* 2(4), 305–317.

Gibbon, P. (2015). How a 17th century philosopher speaks to today's school reformers. *Education Week* 34(37), 23, 28.

Jane Addams quotes. (n.d.). BrainyQuote.com. Retrieved August 7, 2019, from: https://www.brainy quote.com/quotes/jane_addams_133035

Liebovich, B. (2016). Abigail Eliot and Margaret McMillan: Bringing the nursery school to the United States. *Young Children*, 71(2), 92–95.

Lubeck, S. (1996). Deconstructing "child development knowledge" and "teacher preparation." *Early Childhood Research Quarterly*, 11, 147–167.

Mallory, B.L., & New, R. (1994). *Diversity and developmentally appropriate practices: Challenges for early childhood education*. New York, Teachers College Press.

Morrison, G.S. (2018). *Early childhood education today* (14th ed.). Boston, MA: Pearson Publishing.

NAEYC (2009). *NAEYC standards for early childhood education professional preparation: A position statement of the National Association for the Education of Young Children*. Washington, DC: National Association for the Education of Young Children.

Ryan, S., & Grieshaber, S. (2005). Shifting from developmental to postmodern practices in early childhood teacher education. *Journal of Teacher Education*, 56(1), 34–45.

Sample, I. (2009, March). What is this thing we call science? *The Guardian*. Retrieved from https:// www.theguardian.com/science/blog/2009/mar/03/science-definition-council-francis-bacon

Schorr, L.B. (2004). Foreword to the Head Start debates. In Zigler, E. & Styfco, S.J. (Eds.), *The Head Start debates*. Baltimore, MD: Paul H Brookes Publishing.

Science Council (2009). Our definition of science. Retrieved from https://sciencecouncil.org/a bout-science/our-definition-of-science/

Shonkoff, J., & Levitt, P. (2010). Neuroscience and the future of early childhood policy: Moving from why to what and how. *Neuron*, 67(September), 689–691.

Shonkoff, J., & Phillips, D. (Eds.) (2000). *From neurons to neighborhoods: The science of early childhood development*. Board on Children, Youth, and Families, Committee on Integrating the Science of Early

Childhood Development. National Research Council and Institute of Medicine. Washington DC, National Academy Press.

Ulrich, W. (2008). Reflections on reflective practice (2/7): Applied science and expertise, or the art of testing and contesting practical claims. *Ulrich's Bimonthly*, May–June 2008, 6(3), 1–20.

Understanding Science (2019). What is science? Understanding Science. California, University of California Museum of Paleontology. Retrieved from https://undsci.berkeley.edu/lessons/pdfs/what_is_science.pdf

Woodhead, M. (2006). Changing perspectives on early childhood: theory, research and policy. *International Journal of Equity and Innovation in Early Childhood*, 4(2), 1–43.

2

SYSTEMS SCIENCE AND ITS IMPACT ON EARLY CHILDHOOD EDUCATION AND CARE

Sharon L. Kagan and Eva Landsberg

Part I – Background and Organizational Context

Introduction

Historic in subject and scope, this volume recounts how the nascent field of early childhood education and care (ECEC) burgeoned in the mid- to late twentieth and early twenty-first centuries. It will suggest that the emerging field was extruded not only from the needs of parents but from legacies that were imparted to it from an array of hard (i.e., natural) and soft (i.e., social) sciences. Garnering legitimacy in part due to the knowledge obtained from scientific inquiry about young children and their development, ECEC is not only foundationally grounded in developmental science, but has marshalled an array of other sciences (e.g., neuro-, evaluation, systems, implementation) to accelerate its growth and impact. Examining how science has and may continue to impact the field is a both a noble and essential endeavor.

Inherent in this volume's premise (and its promise) are several latent assumptions that have shaped the application of science to the social and humanitarian services in general and to ECEC in particular. First, there is a historic assumption: rooted in developmental science and awash with pedagogical theory, ECEC has been guided top-down by scientists and theorists who posited both nuanced views of human development and specialized learning theories deemed germane to that development. This is to say, generally ECEC has been and is comfortable using theoretical and scientific bases as cornerstones for its foundational premises. Second, slightly preceding the development of ECEC, science itself grew in the twentieth century, becoming characterized by diverse and increasingly sophisticated domains of inquiry, accompanied by highly specialized scientific methodologies. Fueled by growing industrialization, the advent of World Wars, and relatedly America's emergence as a global presence, the advancement of science and scientific inquiry soared to greater prominence. Although legitimized by the establishment of the National Academy of Sciences in 1863, science as the true epicenter for advancement was fueled from diverse investments and societal sectors, including business, industry, philanthropy, higher education, and government. Inherent in this growth was the premise that science can be beneficial to society, promoting health, well-being, and economic growth. Stated differently, the impact of the sciences in ECEC was part of a global zeitgeist that embraced the beneficial effects of science and scientific study.

But a phenomenon so bold does not emerge all at once. Spurred by the Enlightenment, the past centuries have seen a growing embrace of science. Though perhaps originally conceptualized as the purview of elites, there came to be increasing recognition that scientific advancements could improve living conditions for broader society. Both physical and social science needed to be readied for application – in the landmark words of Lindblom and Cohen (1979), it needed to be "made usable." Stemming from this, and important to understand, three premises frame this chapter and the evolution of contemporary science's importance to ECEC. They are: (i) an inherent respect for scientific inquiry as the basis for a field; (ii) an acceptance that science and scientific inquiry can yield considerable benefits to disciplines and populations; and (iii) an understanding that science and practice are not dyadic, but complementary, interdependent, and mutually influencing domains of effort.

Intent and Organization

This chapter focuses on the evolution of systems science and its connection to ECEC. It acknowledges that systems science was extruded at a time when technological advances were rapidly expanding human capabilities and the scope of knowledge available. It also suggests that, like many other sciences, its intellectual corpus evolved over time and was modified by changing contextual needs and opportunities.

In this sense, systems science is both malleable and functional: it affects and reflects its time. With regard to malleability, the application of systems science to ECEC involved the adaptation of theories, which themselves were evolving, to an emerging field. This is to say that both the fields of systems science and ECEC were in flux, rendering the influence of one to the other highly idiosyncratic, non-linear, and remarkably unpredictable. The application of systems science to ECEC was, and is, not a matter of applying a clear formula; rather it is the act of developing many formulas through twists and turns, each one understood and experienced differently by different scholars. It is inherently changeable and hybridized, blending many sciences. Yet in order to enhance functionality, the nascent field of systems science also sought coherence, seeking to render itself maximally usable. Its application to ECEC was forged largely by theorists who lived in the world of practice, whose language was non-technical, and whose goals were to both explain and better reality. Systems thinking in ECEC is rooted in theory but emanates from urgency. As this paper will show, the application of systems science to ECEC is one of remarkable fidelity to context-specific experience; as such, it does not lend itself to easily predictable patterns. Rather, it is an unlikely saga that combines science and serendipity.

This chapter is the story of that unlikely unfolding. It begins with definitions of systems, systems theory, and the emergence of systems science (Part II), and is followed by a review of systems science, tracing its scholarly evolution along with an examination of its parameters and partner theories (Part III). It then turns to a discussion of the evolution of systems science in ECEC, tracing its winding path (Part IV). Part V reviews the effects of ECEC systems work and then speculates on its future.

Part II – Systems, Systems Theory, and the Emergence of Systems Science

Defining Systems

Although the word "system" has become popular in common parlance, agreement on its meaning vanishes with each attempt to delineate its nuances. At the most general level, a system, according to Webster (1970), is a "set of arrangements of things so related or connected as to form a unity or organic whole." This definition suggests that systems are composites of things that are somehow linked together to form an entity that is bigger (and assumedly more productive) than any single individual part. Of course, this is far from a new idea – the Aristotelian idea that the whole is more than the sum of its parts serves as the terra firma upon which more recent definitions have taken shape. Systems scientist Gene Bellinger (2004) describes a system as, "an entity that maintains is existence through the mutual interaction of its parts." Stated most simply, then, systems are a set of things that are united in relationship to one another. Complexity sets in, however, as systems thinking evolves into a science, with each theorist rendering new dimensions to the conception of systems.

The Scientific Roots of Systems Science

Dating back to the early Greeks, the idea of "wholeness" took shape as philosophers contemplated the universe. In seeking to better understand the experienced world, the Greeks embedded their philosophy (and later its descendent, science) in the cosmos, or the natural world order. Conceptualized as intelligible and sequenced, the universe was deemed controllable through rational thought and action. Later, in classical Newtonian physics, the entire cosmos was imagined as a vast machine. This conception of a machine became a general model or paradigm for complex functioning (Mobus & Kalton, 2014). It spawned thinking that later found its way into an array of scientific disciplines, including psychology, sociology, biology, and chemistry, with each investigating the "mechanisms" of various systems. Each considered sub-units (sometimes called sub-systems) that could simultaneously be thought of as independent units in themselves, as well as parts of the whole when amassed with other sub-units. This approach fostered understandings of part–whole relationships, contributing a now accepted premise of systems thinking that suggests that systemic parts or sub-systems have independent life, definition, and function, in addition to their contributions to the whole. Yet, importantly, this classical view of science was still innately reductionist in orientation, seeking to break systems down into their component parts or sub-systems in order to study them more closely. It therefore functioned deductively, from whole to part, on the assumption that a full understanding of each sub-unit would yield a complete understanding of the full system.

Both evolving from and fortifying this cosmic rationale for systems thinking, organic theories took hold. Using both the body and the universe as a metaphor, new theorists accepted the importance of the part to the functioning of the whole, but nonetheless suggested that investigation of individual parts could not provide a complete explanation of the total phenomenon. Rather, they contended, the fundamental character of living things resides in its organization. Individual parts do matter and can be understood as independent entities, but it is the organization (or lack thereof) of these parts that renders coherence and optimal functioning. Stemming from this recognition, another theory of systems thinking took hold, one that legitimated the organization and coordination of parts and processes. This was applied across many fields – Ludwig Von Bertalanffy, a biologist and later one of the founders of systems science, saw that the chief task of biology is to discover the laws of biological systems, both in each sub-system and within the larger system as a whole. Systems are thus defined by their organizational complexity as manifest by the interrelations between many components.

This advancement proved to be the germ of what later would become general systems theory. In sharp contrast to classical science, modern systems science is constructivist in orientation, aiming to build a full understanding of the system through a focus on interactions, not on individual "parts." As Bellinger (2004) contends, in modern systems science, "the interactions of the parts become more relevant to understanding the system than understanding the parts." Systemic thinking values interrelatedness, seeking to understand the nature of the nexus. It has thus been deemed "radical constructivism" by Ernest von Glasersfeld (1981), who posited that systemic thinking breaks with convention in that it focuses on knowledge that does not reflect an objective ontological reality, but instead reflects the way in which the human mind can order and organize of the world, as constituted by our experience. Modern systems thinking thus calls for a re-orientation away from the analytic, mechanistic, linear–causal paradigm of classical sciences and instead introduces "systems" as a new scientific paradigm. As such, systemic thinking was not an easily accepted canon of theory – from the outset, it stood outside the mainstream.

Distinguishing Scientific Thinking, Theory, and Fields

The Evolution of Scientific Thinking, Theory, and Fields

The rise of a systems movement in the mid-twentieth century – characterized by a growing number of scientists and thinkers from various fields who were grappling with systems-oriented ideas – by no means guaranteed the eventual emergence of a codified field of systems science. Formalizing any intellectual zeitgeist or set of errant theories into a recognized science is always a challenge, and particularly so when dealing with a movement as unconventional as systems thinking. Indeed, scholars have long sought to discern the point at which emergent ideas coalesce into a theory, and at what point said theory or theories are well-developed enough to constitute a formalized field. Well beyond the scope of this analysis, a brief review suggests that a movement of scientific thinking is generally manifest in a loosely configured set of ideas, often emanating from diverse disciplines and thinkers, that may share similar themes but rarely converge in full agreement with one another. With time, scientific theories may emerge, taking shape when ideas become codified, when principles are developed, and when relationships among phenomena are established. Though still speculative in intent, scientific theories are typically better-grounded; they garner legitimacy often because they have been upheld by preliminary examination and research. Still, scientists who produce and explore such theories often have diverse institutional "homes," and may work – collaboratively or independently – in a large number of disciplines.

Occasionally, the circumstances might conspire to allow for the emergence of a true scientific field. A complex endeavor, this involves the formalization and classification of study, and in most cases the expression of that study in institutions of higher education. Typically, this occurs when there is a sufficient mass of scholars working on common and/or highly related topics, when there is a codified (though not necessarily consistent) body of literature, and when there is sufficient demand for a formalized institutional presence. Other key factors, including organizational leadership and funding, can also have significant influence on the ultimate success or failure of a nascent field (Reid, 2018).

But the evolution from scientific thinking, to theories, to a legitimized field does not occur in a vacuum. This evolution is heavily contoured by contextual realities. In the past century, for example, the two World Wars and the Cold War occasioned unprecedented levels of investment in the sciences from both governments and the private sector, domestically and internationally. This resulted in a shift to what Bertalanffy (1972) describes as "big science," hallmarked by the rise of large, government- or corporation-sponsored research teams working collectively on the development and application of new procedures and techniques. Motivated by contextual factors (e.g., urgent social and economic crises, social unrest, and globalization), as well as the United States' transition from an industrial economy to a post-industrial, knowledge-based economy, the scientific enterprise grew (Klir, 1991). With it, theorists began to consider new paradigms and more diverse applications for their work.

This notable shift provided especially fertile ground for systems science. As systems scientist George Klir (1991) describes, in the industrial era, knowledge was generated largely via "random" inventions, while in the post-industrial era more innovation is derived from the codification, synthesis, and integration of theoretical knowledge. The strategic resources of the post-industrial society are information and theoretical knowledge; with the rise of computer technology and the cybernetic movement, there is a fresh focus on discerning linkages and interactions. The sheer complexity and near-infinite amount of data and information now available demands a more systematic approach to science, one with less emphasis on managing *things* (characteristic of the industrial era) and far more on managing *complexity* (Banathy, 1995).

The Evolution of Systems Thinking, Theory, and Fields

As discussed previously, systems thinking emerged in a loosely configured systems movement in the twentieth century. This movement boasted a growing number of scholars from diverse disciplines, who in time became loosely associated by their common interests in thinking systemically and strategically about concepts, principles, and methods that could be applied across disciplines. Foremost among these men were Bertalanffy (biology), Kenneth Boulding (economics), Ralph Gerard (physiology), and Anatol Rapoport (mathematical biology), along with W. Ross Ashby (cybernetics) (Klir, 1991). Notably, though their scientific foundations were quite distinct, they found common ground in their concern with systemhood (i.e., the way things are organized and the properties of systems) rather than thinghood (i.e., what they are). Thus, the systems movement was at its very core interdisciplinary. As Mobus and Kalton (2014) would later describe:

> Systems science provides a somewhat unique mode of inquiry in revealing not just how one kind of system, say a biological system, works, but rather how all kinds of systems work. That is, it looks at what is common across all kinds of systems in terms of form and function.

Those early leaders of the systems movement thus sought to explore "those properties of systems and associated problems that emanate from the general notion of systemhood" (Klir, 1991).

Ultimately, the systems movement would produce a wide number of theories dealing with the nature and interconnections of systems. Foremost among these is General Systems Theory, now considered to be a foundational principle of systems science and elaborated subsequently in this chapter. Gradually, as agreement coalesced around certain systemic ideas and collaboration increased between systems scholars, systems thinking began to make the transformation from being an intellectual movement to being a defined field of science. A notable component of this shift was the inclusion of "systems science" in university curricula, which ultimately evolved into the establishment of permanent systems science-focused entities within institutions of higher education. Some of these entities stand alone as individual departments – the Systems Science Department at SUNY Binghamton, for instance – while others, such as the Survey Research Center and the Group Dynamics Center at the University of Michigan, are interdepartmental in nature (Klir, 1991). While these entities are typically aligned with the fields of engineering, mathematics, and computer technology, they nonetheless produce scholarship that transcends diverse disciplines, both in the physical and social sciences.

Moreover, as systems science took hold as a recognized field of study, professional societies emerged. By 1980, there were three such societies in the field of "systems research," which ultimately came together to found the International Federation for Systems Research (IFSR) in Vienna. Its mission was to advance systems science across an array of theoretical bases including "cybernetics and its special fields, the investigation of its applications on technical, economic and social problems, [and] its relations to problems studied in the natural sciences and the humanities" (IFSR, n.d.). As controversial and unconventional as systems science appeared at the outset, by the late twentieth century it was a young but established field of study, one that had prominence and acceptance internationally, and one that would soon come to have profound influence on the field of ECEC.

Part III – Systems Science: Parameters and Partner Theories

Anticipating a detailed discussion of systems science as it applies to ECEC, which follows in the next section, this section focuses on the parameters and principles of systems science and situates it with other theoretical movements that have also helped shape ECEC.

Parameters of Systems Science

Having traced the evolution of systems science in the prior section, it is important to make its parameters transparent. Building on some of themes enunciated earlier, the field of systems science is generally characterized by the following:

Universal Applicability

Like all sciences, systems science is based on numerous theoretical understandings, perhaps the foremost of which is Bertalanffy's General Systems Theory (GST). Widely accepted, this theory declares systems science to be the study of "models, principles, and laws … which apply to generalized systems irrespective of their particular kind" (Rousseau, 2015; Bertalanffy, 1972). In other words, systems science applies to "systems in general," whether in the physical or social sciences. Bertalanffy's GST thus establishes an assumption of the universality of systems – that there are certain commonalities shared by all systems, across all disciplines. As a result, systems science is considered by some to be a *meta*science that embraces and can be applied meaningfully across multiple sciences concurrently (Mobus & Kalton, 2014).

Synergistic Interrelations

As previously discussed, classical science worked to isolate elements of the observed universe, with the expectation that putting them together would render a more complete understanding of the full system. Systems science embraces the necessity of understanding interrelations, so that a system is not merely a collection of aggregated pieces, but also relies on the connections between the pieces. Mobus and Kalton (2014) said it well: "The elements in a system do not simply pool their functions to add up to the behavior of the whole; they rather perform in synergy with one another so that the behavior of each is critically shaped and informed by its relation to the whole."

Environmental Interactions

Systems are generally characterized as either open or closed. An open system allows for interactions between the components of the system and the environment, while a closed system remains isolated from its environment. As might be expected, social sciences generally use open systems as their models, since they have "intensive co-evolutionary interactions with their environment." The feedback these systems receive from their environment often results in the need for new goals, new perspectives, new functions, and even complete systems-level change (Banathy, 1995).

Multiple Purposes

A unitary system is one characterized by a single goal, one which is often quite stable and transparent (Banathy, 1995). Most systems, however, are pluralistic systems, wherein there are many, sometimes conflicting, goals. Often, these conflicts are unavoidable results of systemic complexity. Systems science calls for these multiple and conflicting goals to be acknowledged, and their effect on overall systemic functioning to be considered through careful analysis.

Normative Change and Complexity

Open systems are rooted in multiple interactions with the environment; as such the nature of the interactions changes and this change is normative and to be expected. Systems science accepts change and complexity as an inherent parameter, one to be studied and understood as an integral component of understanding the variability and functions of systems. It expects fluctuations in relationships, deeming complexity over time and across contexts as normal.

Taken together, these five parameters describe some basic tenets of systems science. As such, they not only predict the nature of scientific inquiry that the field evokes, but render it a science that is well-suited to understanding an open, chaotic, complex, multi-purposed, synergistic, and environmentally interactive field like early childhood. With no single funding source, with elements of ECEC bumping against each other and acting synergistically even without intent, with its community- and environmental-nestedness, and with numerous recalcitrant challenges which often transcend diverse applications, ECEC could be the poster child for applied systems work.

Partner Theoretical Frameworks

As robust as systems thinking and systems science is for the study of ECEC, several allied or partner theories are also essential to understand. Among these, four will be highlighted here: functionalist theory, neo-institutional theory, general organizational theory, and network theory. At their core, these theories share properties with systems science because they emanated from a similar scholarly zeitgeist in the mid-to-late twentieth century. Fueled by the rise of technology and globalization, these allied frameworks attempt to deal with the expansion of information, and with it, complexity. Each acknowledges the centrality of environmental context and recognizes that change is not linear, but highly mutable. Like systems theory, these partner frameworks acknowledge the role of systemic interactions, although each focalizes systemic elements somewhat differently. For some scholars, these theories represent an evolution of systems thinking; for others, they are variants to it. They are presented here because they represent viable ways of "holding" contemporary systems thinking and are often used to explain phenomena affecting ECEC domestically and globally.

Functionalist Theory

Rooted in the same mid-twentieth century systems movement that spawned the field of systems science, functionalism emerged as a more applied branch of systems theory. Functionalists operated in the social sciences sphere, examining the interrelatedness of different aspects of social relationships and culture from a systemic perspective. Scholars including Simon, Bronislaw, Malinowski, and Radcliffe-Brown conceived of organizations as entities composed of subsystems, which were malleable in response to externally imposed conditions. Functionalist theories therefore made important intellectual contributions to systems work in the social sciences, particularly around issues of stability, order, and change. Rather than focusing on "prediction and authoritative control," they emphasized that systems were subject to constant change, not necessarily as a result of disequilibrium or problems within the systems themselves, but as the result of changes to the external environment (Kagan & Neville, 1993).

Neo-institutional Theory

Formulated in the 1970s and 1980s, neo-institutional theory proposed that structural properties shape and are shaped by technical and resource demands and dependencies. These variables are contoured by institutional forces such as rational myths, professional expectations and requirements, public opinion, and the law. Similar to "open systems" thinking, neo-institutional theory sees organizations as deeply embedded in their social and political environments, which in turn are fluid and subject to change. As such, neo-institutional theory accepts the complexities that are associated with systems theory, and seeks to discern and expose institutional regularities, practices, and structures as either reflections of or responses to rules, beliefs, and conventions that are built into the wider environment. Neo-institutional theory rejects the notion of institutions as fixed and static entities (e.g., rationalist theories); rather, it notes that institutions, like people, change in response to the temporal and cultural zeitgeist (e.g., adaptist theories). Neo-institutional theory, as an applied variant of systems science, accepts complexities and further seeks to account for them as a way of both accepting and predicting change.

General Organizational Theory

Closely aligned with systems theory, organizational theory seeks to describe, explain, and predict complex interactions between organizations and their external environments. Rooted in management, sociology, and psychology, organizational theorists seek to remedy complex organizational challenges by nesting them in a systems perspective. Thus, rather than attempting to break down issues into discrete problems (reductionism), they seek to explain complex phenomena by looking at the interactions among players, parties, and challenges. With an emphasis on description, rather than prescription, organizational theorists commence their work with the assumption that improving the quality of organizations is possible and can be best accomplished by examining interactions between the individual and the organization. In this sense, a systems perspective is inevitable, as the phenomena of organizational life are deeply embedded in the structures and patterns that comprise the system. In advancing this stance, organizational theorists believe there are shared organizational principles and practices that exist across a wide variety of settings, though they acknowledge that cultural context plays an important role in influencing organizational trajectories and values. In the realm of ECEC, organizational theorists may focus on individual programs or centers, or systems/boundary-spanning entities such as councils or agencies.

Network Theory

Given the growth of cooperative arrangements among organizations and the related advent of action hubs, networks are not only becoming more common as delivery mechanisms, but, given their more loosely configured organization, they are also becoming a topic of burgeoning interest theoretically. Though often less structured than formal systems, networks share many properties with open, pluralistic systems. Namely, networks are deeply embedded in the environment, are quite malleable, and often function to support organizations with divergent purposes, structures, and modes of organization.

Network theory examines the way the elements of a network interact and connect, beginning from an individual level and progressing to examine large-scale questions of social order. In other words, it aims to explain how human beings – each of whom is embedded in a thick web of relationships with those around them – come together to form enduring, functional organizations or societies. In so doing, it has helped generate explanations of phenomena across the social sciences. In recent years, it has been applied to fields such as national security, public health, communications, and management consulting. When applied to service-oriented fields such as ECEC, network theory posits that each service is embedded in its own network, and seeks to form connections that allow for easy communication and cooperation between the "nodes" (e.g., individual service providers). Conceptually aligned with systems thinking, network theory emphasizes the importance of relationships and structure, operating under the assumption that networks can be formed in many different ways, and that the forms they take can have significant implications for effectiveness, efficiency, and quality (Borgatti, Mehra, Brass, & Labianca, 2009).

Systems science and its allied theories represent evolutions in thinking, which are sure to unfold beyond their present formulations. Yet, for those concerned about the status of early childhood education, systems thinking has proved to be a useful rubric within which to better understand and advance the practice of ECEC. Just why this is the case and what makes ECEC so ripe for systems work will be examined next.

Part IV – The Evolution of Systems Science in ECEC

The Rationale for Systems Science in ECEC

Embedded in a social context that values the primacy and privacy of the family, America's public policies for children and families had a slow start. Predicated on strong beliefs that government should intervene only when families fail, and that caring for children, especially young children, was the joy and responsibility of the family, America's public ethos historically inhibited the development of robust and consistent policies for young children. This value proposition began to change, however, in the mid-twentieth century, in part as a result of the civil rights and women's movements. Augmented by more recent scientific advancements, many of them detailed in other chapters in this volume, public engagement in supporting the lives of young children and their families has notably increased in the United States over the past decades.

But this rising interest in young children has not been smooth or consistent, nor is it clearly lodged within any single level of government, single sector, or any single comprehensive policy. Indeed, the evolution of services to young children has been hallmarked by fits and starts that left an inchoate array of uncoordinated, often inequitably distributed, services. As noted previously, this began to emerge in earnest in the twentieth century, when America's embroilment in international and economic crises including the World Wars and the Great Depression summoned men and women into public service, thus provoking the need for child care. With typical American zeal and ingenuity, comprehensive systems of early care were developed and implemented, only to be dismantled at the end of the public crises. Children's policies were seen as handmaidens to larger social causes; they were easily mounted and summarily eliminated according to social agendas and needs that were not predicated on children. In its wake, this episodic approach to policy left a legacy of programmatic vulnerability, unprofessional status for teachers, and inconsistent expectations for government involvement in ECEC.

Continuing to intervene at times of national crisis, in 1965 the federal government established its landmark Head Start program, which changed the zeitgeist regarding the need for and operationalization of services to young children and their families. Buffeted by a broad-based constituency and a reliably supportive Congress, Head Start stands unique in its scope and durability. Although for many years it was the primary public program that supported early learning and development, as women marched into the workplace and data attesting to the value of quality early childhood programs landed on desks of America's elite, new programs and services proliferated. Highly diverse, these varied in audience (children, parents, both), intention (education, health, parenting support), auspice (schools, community-based settings), and sponsorship (public or private). The long-held American values of entrepreneurialism and localism framed this chaotic growth, resulting in a proliferation of uncoordinated services that were not only ill-organized and complicated (if not inequitable) to access, but were also lacking the supports needed to render them of requisite quality to achieve the impressive outcomes expected of them.

Amidst this chaos, early childhood programs came and went, flourished and floundered. Sometimes brought to life by enthusiastic governors, mayors, or first ladies (there were few, if any, first spouses, then), programs were often summarily eliminated at the behest of a subsequently victorious politician with a different vision or political affiliation. When programs were sustained over time, they were often layered onto an early childhood landscape that was ill-prepared to greet them. Shortages of adequately trained staff fueled competition and "staff-stealing" among programs. Inadequate state licensing personnel and ill-formed monitoring systems overlooked poor performing programs, suppressing the effects of the then-limited efforts to improve quality. Moreover, burgeoning need gave rise to a robust private sector, with large numbers of for-profit programs, incentivized by economic gain, surging to the fore. "Underground" services also blossomed, with much child care bartered among family, friends, and neighbors.

Therefore, though services for young children proliferated with each new research study and public information campaign proclaiming their importance to children's and societies' development, they too often lacked cohesion and continuity, and were often bereft of quality and equality of access. Though welcome theoretically, functionally the groundswell of efforts fell on the most fragile of ecosystems, a social platform that was ill-prepared to handle rapid expansion. Moreover, amidst a rising accountability movement that demanded efficiency and positive outcomes for young children and families, there was little guarantee that the promise of early childhood could or would be realized. By the late twentieth century, recognition of this fractious situation gave rise to calls that the field was stuck and needed re-visioning – a new way to be conceptualized, planned for, and understood.

Enacting a Systems Orientation: The Formative Years

Paradoxically reflecting both a divergence from the past and a reflection of it, the 1980s and 1990s hallmarked major reform in ECEC. Characteristic of a new era, leaders in this period parted with the past to guide major social change. Such change did not emanate from external forces, like a national crisis. Rather, the "crisis" that propelled progress came from within: ECEC leaders recognized that the moribund non-system of scattershot programs and services was having latent negative consequences. However well-intentioned the ECEC policy efforts, services on the ground lacked comprehensive planning, operational quality guidances, adequately prepared personnel, and appropriate approaches to monitoring and accountability. Above all, access for those who were the intended and most notable beneficiaries was not equitably distributed. Due to poor planning, some neighborhoods were stacked with services with spaces going empty, while others faced a notable paucity. In short, there was a felt need to address the effects of rapid and unsystematic expansion – in short, to take a policy or systemic stance. Reflecting the past and like so many prior pedagogical innovations for young children (e.g., Montessori, cooperative nursery schools), the press for policy innovation swelled more from felt need than from the scientific community. To bring coherence to an unruly field, leaders began to adopt a systems orientation.

Notably, however, those in the field of ECEC were not the only social service leaders influenced by the rise of the mid-century systems movement. Bronfenbrenner, for example, applied systems thinking to his ecological model of development; Bowen applied systems science to family theory; and Banathy (1995) called for an adoption of systems thinking for education. Wrote Banathy:

> Rephrasing Einstein, our current problématique of education cannot be solved from the same consciousness that created it. What created it – in addition to a persistent resistance to change – is the still prevailing fragmented, disjointed, and piecemeal approach in educational inquiry. Our challenge is to "learn to see the world anew," to develop a new mind set, and to acquire new thinking that is based on a systems view of the world.

Clearly, the press for systems thinking in ECEC was not solely the purview of early childhood leaders – it had many roots.

Early Action in the Private Sector

The groundswell for systemic action, like the problems it sought to rectify, was messy. And it was quiet. There was little fanfare, little coherent planning. It did not begin with a stunning proclamation of a huge social problem, although a 1990 National Research Council of the National Academy of Sciences report on child care did document that there were a staggering 31 programs and 11 federal agencies serving young children (Hayes, Palmer, & Zaslow, 1990). It did not emerge as a Nobel prize-level scientific breakthrough; there were no startling new discoveries. Rather, taking twisting paths, the movement toward systems work took hold in the heads and hearts of individuals who had practical knowledge about both ECEC services and young children. Not always working in collaboration, these individuals occupied roles in the public and private sectors, in governments, universities, and foundations. What they shared was a pressing concern that all was not right with the cascade of earnest policy efforts that were being mounted on behalf of young children.

Among the first to identify the challenges was a beloved civil servant, Jule Sugarman, who oversaw government programs for young children. In 1991, he published the notable *Building Early Childhood Systems* handbook, which presented embryonic and prescient ideas about what constituted an early childhood system. Defining a system as "a set of arrangements under which individual programs and activities work with one another," he proffered a number of effectiveness parameters, including the development of common standards, cross-program planning, and increased government support. Indeed, so strong was his vision that he boldly called for coordination among programs, over and above their continued creation. Sugarman's visionary efforts were thus seminal in documenting—though not necessarily altering—the problem.

In calling for the creation of linkages or a set of new relationships among programs, Sugarman (1991) assumed that all the pieces existed, and that the problem lay in their isolated arrangement. In contrast, Kagan and her colleagues noted that the field was stuck, but attributed this not to the arrangement of the extant services, but to a lack of understanding of and support for the pieces. Describing these missing elements as the essential functions of an early care and education system (Kagan & The Essential Functions and Change Strategies Task Force, 1993), early childhood leaders laid the ground work for a bold "re-think." From this emerged the collaborative Quality 2000 effort, which pulled together the thinking of over 350 individuals in the field as well as much scholarship from related fields on systems thinking. The final report of the effort, *Not by Chance*, advanced a number of recommendations and identified five important – and too often, missing – elements of ECEC system infrastructure: (i) parent information and engagement; (ii) professional development and licensing; (iii) facility licensing, enforcement and program accreditation; (iv) funding, financing; and (v) governance, planning and accountability (Kagan & Cohen, 1997). The report, anchored in systems thinking, noted that each element needed to be present in order for a coherent system of early care and education to exist.

The report also specified that in order for a system to exist, attention needed to be focused on both the programs (i.e., direct services), as well as the infrastructure that supports the programs. In presenting this work publicly, a formula was posited declaring that an ECEC system = programs + infrastructure. Later the infrastructure was elaborated to include a total of eight infrastructural elements and another formula, 8−1=0, was advanced, reflecting the concept that a system could not exist without each and every one of the components. Later, a graphic that depicted "flowers and gears" gained currency as a means of rendering complex systems thinking more concrete. It suggested that the flowers were the programs that were being planted, but without sufficient attention to the soil (i.e., the infrastructure), the flowers would die. To further concretize the relationship, the infrastructure was depicted as having gears, all of which were needed to create the ideal conditions in which an early childhood system would thrive (Kagan & Cohen, 1996).

Gradually, systems thinking gained currency throughout the ECEC field, with thought leadership extending its purview to include diverse systems that affect young children, including health, education, family support, and services for children with disabilities (Bruner, n.d.). So pronounced was the emerging commitment to systems work that, in May of 2002, a group of funders came together to create the Early Childhood Funders Collaborative (ECFC) with the goal of addressing the schisms that existed in service provision for young children and their families. Dedicated to addressing the reality that programs, policies, and services often operate in isolation, at cross purposes, or without enough resources, the ECFC continues to exist today and maintains a fervent commitment to advancing the field. Through its robust funding, the ECFC has been a notable contributor to significantly advancing systems thinking and systems progress in ECEC.

Early Action in the Public Sector

Along with these early private-sector forays into ECEC systems thinking, a surprising number of public initiatives, many of them emanating from the federal level, took hold. As early as

1975, when the *Education for All Handicapped Children Act* (Public Law 94–142) was passed, there was an acknowledgement that children with disabilities had a right to an education, and that such efforts needed to be linked with health and other supportive services. Moreover, in light of the atrocities that happened to some children in institutionalized care, the legislation brought to the fore the need for not simply institutional reform, but systemic reform. When amendments to this law were passed in 1986 (P.L. 99–457), a new federal discretionary program was created to help states develop and implement coordinated systems for handicapped children. Distinct from 94–142, this act was explicit in the attention it accorded service integration, requiring that states transform their fragmented delivery systems into comprehensive, multidisciplinary, and coordinated systems (Gallagher, Harbin, Thomas, Clifford, & Wenger, 1988). Elaborated later on in *Individuals with Disabilities Education Act* (IDEA), intentions to build systems were clearly stipulated as part and parcel of these efforts so that, for example, the 2004 reauthorization of IDEA mandated that states establish an interagency coordinating council to advise them on the implementation of the law's Part C program for infants and toddlers.

Health and mental health policies also made strong and early headway in systems work. Through a series of efforts launched by the Maternal and Child Health Bureau (MCHB) of the Health Resources and Services Administration, a network of resource centers was established to advance community-centered, client-based, comprehensive early identification and collaborative services. In 1989, through the *Omnibus Budget Reconciliation Act*, funds were provided to promote service integration and systems building. This legislation also required linkages with the Early and Periodic Screening, Diagnosis, and Treatment effort, that itself sought to align services and systems so as to more effectively serve young children. The MCHB's Early Childhood Comprehensive (ECC) Systems Initiative has provided grants to almost all states (47 states and 5 jurisdictions) to support systems development. Somewhat unique in scope and reach, this initiative supports efforts to build partnerships between interrelated and interdependent agencies/organizations representing physical and mental health, social services, families and caregivers, and early childhood education to develop seamless systems of care for children from birth to kindergarten entry. As such, grantees are required to focus simultaneously on systems building across several different sectors, including ECEC, health care, mental health, family support, and parenting education.

The public early childhood field took numerous steps to launch coordinative efforts, with Head Start often at the helm. In 1990, for example, the federal government began funding state Head Start Collaboration Offices with the express purpose of creating systems that link Head Start and the other child development-serving entities. Always deeply ensconced in local communities, Head Start, by dint of its federal-to-local funding and governance structure, does not parallel the funding apparatuses of other programs, making systems-building challenging. Recognizing this, the federal government, supported with some state contributions, worked early on toward systems development. And in addition to these federal efforts, many municipalities established coordinating mechanisms and efforts. Notable among the early efforts, North Carolina's Smart Start was created in 1993 by collaboratively linking the public and private sectors to create a comprehensive system of services that also, via the North Carolina Partnership for Children, Inc. and local partnerships, works at all governance levels.

Overall, the formative years of systems work in ECEC took hold somewhat idiosyncratically in both the private and public sectors. Several things characterize this era. First, there was clear manifestation of a shift in thinking from merely a programmatic to a more systemic orientation is both sectors. Second, this thinking was manifested in disparate disciplinary systems, notably health, mental health, education, and early childhood/early learning. Third, there was an abundance of trial and error efforts that were designed to promote systems thinking and systems action. Fourth, few of the efforts seemed to be guided by a concrete theory of change or even theory of action. Fifth, there were few concrete efforts to evaluate these efforts, either in terms of implementation or their theorized outcomes. And finally, and most germane to this paper, they tended to have all emerged from felt need, rather than as a direct outgrowth of the nascent field of systems science. In this sense, ECEC appears to have primarily embraced the systems thinking *zeitgeist* of the era, employing a systemic orientation – though not specific systems science principles – to respond to the urgent needs of the field.

Enacting a Systems Orientation: Recent Efforts in the Field

Continuing the trends launched in the formative period, recent years have been hallmarked by more organized efforts, both public and private, that are grounded in a systems orientation. Notable is the fact that as systems work has emerged, although one sector may take the lead, dyadic relationships have formed, so that work produced in one sector is informed by and contributes to the other. Progress has been made in the elaboration of many of the elements of the systemic infrastructure, essentially creating a series of sub-systems related to professional development, data and accountability, and governance. Further, although presented below in two distinct categories according to their primary impetus and/or sponsor, the sector segmentation characteristic of earlier years is dissipating as collaborative working relationships emerge. Thus, although there has not yet been a full coalescence of practice and theory, and systemic theory still tends to lag behind systems practice in ECEC, field-based work is noteworthy.

Recent ECEC Systems Work in the Private Sector

With bold intentions, an independent group of individuals interested in systems work voluntarily came together in 2005 to form the Early Childhood Systems Working Group (ECSWG). Composed of individuals and representatives of numerous organizations, including the Build Initiative, the ECSWG sought to support states in implementing systems that would promote an integrated continuum of services across early learning and development, health, and family leadership and support. In 2006, the group developed an early childhood systems framework, commonly known as the four ovals: early learning, health, mental health and nutrition, family support, and special needs/early intervention. Predicated on theoretical work and codified by Charles Bruner (n.d.) in *A Framework for State Leadership and Action in Building the Components of an Early Childhood System*, the ovals were used for a number of years. In 2015, they were reconstituted to include just three ovals, reflecting a desire to promote the integration of all children with special developmental needs and challenges into each oval, and to emphasize the active role of families. The reformulation has thriving children and families at its core, reflecting an accountability trend discussed later on. Surrounding the three ovals are six functions, each of which supports the three ovals: (i) define and coordinate leadership; (ii) finance strategically; (iii) enhance and align standards; (iv) create and support improvement strategies; (v) ensure accountability; and (vi) recruit and engage stakeholders. The ECSWG has also produced an array of related resources, including a systems planning tool.

Aligned with this work, other efforts have also taken hold to promote systems work in the states. In 2006, for example, the National Governors Association awarded grants to three states to fund in-state meetings and technical support, followed by smaller grants to 12 states to hold early childhood summits, all of which were designed to link elements of the system in service to better outcomes for children and families. Moreover, such work is being chronicled for policymakers and the public. For instance, the Bipartisan Policy Center recently released an analysis of integrated services in ECEC (2018), as has the Education Commission of the States (Atchison & Diffey, 2018). Both efforts document field-based efforts to create more integrated, systemic services.

Indeed, much ground-breaking work has taken place to affirm and advance ECEC systems building. Often taking the form of technical assistance, innovative and well-researched reports, and applied work in the field, efforts are often undertaken to bring together thought leaders for productive meetings and consultations. In April of 2007, for example, the Build Initiative held a symposium for 60 people on ECEC systems efforts, which set the stage for more events that followed. With the goal of aligning goals and coordinating actions across programs with different political cultures, a more nuanced contextual orientation has come to characterize this work. In particular, there is a growing emphasis on the nature of context, and a more purposeful integration of the role of culture and values into systems work, rendering it simultaneously more adaptive and more complex. Concerns had mounted that without such an orientation, systems work would die lifeless on the shelf. This is a clear example of the dyadic nature of theory and practice, providing a strong tilt to the importance of the practice base.

Systemic work in ECEC has also been influenced by the rise of the accountability movement. At the most basic level, the drive for early learning and development standards that characterize what young children should know and be able to do provides the basis for pedagogical reform in ECEC (Kagan, 2012). Such efforts are not typically labeled as systemic reform – but in fact, by specifying standards that transcend programs, a common approach to early learning provides a uniform and systematic foundations for all programs. The standards provide a backbone for numerous applications, including curriculum development, professional preparation for those working with children, parenting education curricula, and assessment development. In basing all these elements that ultimately affect young children on standards, systemic integrity is created. No longer are teachers being prepared on content they will not teach, no longer are children being assessed on content to which they have not been exposed, and no longer are parents creating false expectations for their children's development. As a foundation, early learning and development standards provide the common platform not only across programs but across diverse pedagogical applications.

Accountability has also set in motion the development of comprehensive data systems that seek to collect and report uniform information across ECEC programs. Based on the understanding that data can drive improvement in services, results-based planning has been embraced so that new services and policies can be more effectively organized and implemented. To better understand the current status of children, families, neighborhoods, and communities, it is important that data are routinely collected and then compiled into effective data systems, which can in turn provide the evidentiary basis for social and policy improvement. In the field of early childhood, data systems even *within* individual programs and services (e.g., Head Start, child care, mental health) are somewhat limited; their ability to be linked *across* services is virtually non-existent. To address these issues, an Early Childhood Data Collaborative, a consortium of seven national organizations, was formed in 2009 (Gruendel & Stedron, 2012). As a first step, the group documented systemic data efforts in the states, noting those that had more advanced systems (Early Childhood Data Collaborative, 2010) and could provide information not simply on enrollment counts, but on unduplicated enrollment counts, costs per child across programs, and outcomes achieved in light of dollars invested. Using this information, the group – backed by grants from the federal government – advanced early childhood data systems work throughout the United States. In so doing, they have not only dealt with tough issues regarding confidentiality, but have been instrumental in advancing a number of practice innovations including the development of comprehensive data audit processes and of identifying numbers for children, the workforce, and programs. Moreover, they have advanced the idea of developing a comprehensive data system for all early childhood services.

Recent ECEC Systems Efforts in the Public Sector

Perhaps even more startling than recent advances in the private sector is a set of hallmark efforts in the public sector that has fueled systems work. Though taking different forms and having different functions, governance entities can and do serve as another critical platform that brings disparate elements of the ECEC field together (Kagan & Gomez, 2015). Their efforts often begin as collaborative or planning entities but gradually become formalized, sometimes as new agencies or ministries (as in Massachusetts) and sometimes as interagency entities (Kauerz & Kagan, 2012). Often they take on diverse authorities and accountabilities, but in all cases they provide systemic "glue" that binds diverse programs and services.

Acknowledging the challenges inherent in creating an ECEC system, the 2007 Head Start Reauthorization Act mandated that governors in each state designate an Early Childhood Advisory Council to develop a coordinated system of early childhood education and care. Although unfunded for a period of time, this mandate reflected the importance of such systemically oriented councils. In 2008, six states received federal grants to support systems-building and child wellness through Project LAUNCH (Linking Actions for Unmet Needs in Children's Health). The same year, Congress authorized the creation of a grant program to support the development of Early Childhood Education Professional Development and Career Task Forces. Highly important, these efforts have continued through the years, so that as recently as July 2016, the Health Resources and Services Administration awarded 12 five-year ECCS Impact grants and one Early Childhood Comprehensive Systems Collaborative Innovation and Improvement Network Coordination Center Cooperative Agreement to foster systems work.

As important as all these efforts were and are, they are somewhat diminished in scale and scope by the legion Early Learning Challenge Fund (ELCF) that arrived in February 2009 as a part of the American Recovery and Reinvestment Act. With an aim to stimulate the entire United States economy, the federal government allocated collaborative responsibility for the ELCF to the US Departments of Education and Health and Human Services. Together, the two departments planned and funded a set of highly competitive grants that provided significant funding for a limited set of states to advance their work and efforts for young children. The goal was to use systemic approaches to drive results-oriented, standards-based reform across programs, foster programmatic improvement, and improve kindergarten-readiness rates. This effort marked the first major and incentivized systemic effort in the field, in that its funding ($8 billion) was on a scale heretofore unseen.

Given the stipulations of the ELCF and the fact that this was one-time funding, the resources were used for a variety of important efforts that helped advance systems work. Notable among them were investments in the quality rating and improvement systems (QRIS) which had begun to take hold in the late 1990s, and by 2009 were operative in 26 states (Tout et al., 2010). Research based on empirical studies that attested to the relationship between program quality and children's development (Campbell & Ramey, 1994; Schweinhart, Barnes, Weikart, Barnett, & Epstein, 1993) provided the impetus for the development of new and systematic approaches for program quality enhancement. Beyond providing information to policymakers, the QRIS are designed to help program providers discern areas for improvement and then use resources (now more readily available in ELCF states) to make discernible quality improvement. Through well-defined and typically publically available rating systems, ECEC quality was also rendered more transparent to the public. Originally designed as an accountability system that would infuse a uniform level of quality expectations across programs, QRIS gradually expanded to serve as a vehicle for linking diverse programs and funding streams in a quest for common quality standards and benchmarks. Discussed elsewhere in the literature (Schaack, Tarrant, Boller, & Tout, 2012), QRIS, although different in each state, have both operationalized and popularized systems thinking and action in ECEC.

The consequences of these broad efforts in the private and public sectors have helped create a new systems zeitgeist in the first decade of the twenty-first century. Once regarded as ancillary and highly speculative, systems initiatives have gradually become more central to the early childhood agenda. Indeed, accelerated by work of foundation philanthropists systems thinking in ECEC now abounds. Yet however widely embraced the idea of systems-building has become, its actual accomplishments seem still somewhat vague; they remain understudied and undocumented. That ECEC systems work nonetheless continues to garner attention suggests that it is driven strongly by a practical imperative even though its effects are not documented, a trajectory that may not parallel the use of other sciences in the ECEC field.

Theorizing ECEC Systems: Recent Advances and Pressing Needs

Recent Advances in ECEC Systems Thinking

Having noted a disconnect between systems science theories and on-the-ground ECEC systems work, there have nonetheless been conceptual and theoretic advances that have indeed impacted ECEC systems work. By no means a codified field of science, the work of scholars from diverse organizations is aggregating to form a respectable body of literature on ECEC systems. A great deal of recent systems-oriented ECEC research focuses on governance and the need to create a home for systems work in the states. Gomez (2015) has extensively examined different approaches to governance and has suggested that fresh conceptual framing would advance the field. Building on complex adaptive systems work, she posits that if government is to be an elixir of systems work, it must focus not only on structural properties, but on functional ones, as well. Citing efforts that have taken hold across the country, Dichter (2015) writes convincingly about the importance of structure, providing a wealth of understanding regarding the implementation and scaling of systems efforts. Yoshikawa, Wuermli, Raikes, Kim, and Kabay (2018) suggest that systems work is the necessary ingredient to foster both scalability and sustainability of ECEC efforts. Combining a wealth of theoretical knowledge, Regenstein (2015) has proffered a typology of governance models that, if implemented, would accelerate systems thinking and systems actions.

Pressing Needs in ECEC Systems Thinking

Despite these notable efforts, there are emerging challenges facing those who are concerned with advancing ECEC systems thinking and systems work. All related to the advent of the accountability movement, three areas will be highlighted: (i) increasing demands and efforts to evaluate systems work; (ii) the emergence of "theory of change" efforts to render a cogent methodology to such evaluations; and (iii) a re-clarification of the need for and content of systemic outcomes.

Increasing Demands for and Efforts to Evaluate ECEC Systems Work

Given pressing demands to use increased resources on direct services for children, systemic work is hard to justify, despite the efforts cited above. Justifying the diversion of funds from services for children to systems work is made more difficult by the lack of a solid research base that affirms the cost-effectiveness of systems investments. Indeed, however much it has become legitimized, ECEC systems work begs for evaluation as a means to rationalize its existence and prove its efficacy. Under the aegis of the BUILD Initiative, a study of systems was undertaken to address this need. Designed to examine how systems were unfolding across the states, the resulting e-book, *Rising to the Challenge: Building Effective Systems for Young Children and Families* (Jordan & King, 2015), examined how different ECEC governance structures emerged, and the role of governors' offices in launching and sustaining them. The study looked at how such systems efforts unfolded, examining the number and nature of funding streams, the existence of consolidated governance entities, and the strategies inherent in aligning goals and coordinating actions across programs with different political histories and cultures.

On a global level, systems initiatives are not uncharted territory. As part of a broader initiative funded by the National Center on Education and the Economy, Neuman, Roth, and Kagan (in press) analyzed 16 multi-country studies that focused on one or more elements of the ECEC system infrastructure. When considered in the aggregate, these studies covered an impressive breadth of countries, attesting to the widespread embrace of ECEC systems thinking across the globe. Yet importantly, the studies examined by Neuman, Roth, and Kagan revealed that international systems varied considerably, and that the presence of systems work did not necessarily provide a panacea to address the range of complex challenges that governments sought to ameliorate.

Moreover, these studies – sometimes called evaluations of systems building, systems change, or systems reform – face large methodological challenges that prevented data aggregation. For example, the international studies were often conducted at different points in the evolution of a system, and they took place in contexts that were not only very different, but that embraced systems thinking very differently. Further, like any evaluation of a complex, multifaceted phenomenon, the studies were somewhat challenged in the ways they conceptualized and operationalized systemic work. More often than not, the studies succeeded in describing the extant systemic elements and delineating the complexities associated with launching systems, but did not focus on evaluating their impact or outcomes.

The Emergence of "Theory of Change" Efforts to Render a Cogent Methodology to Such Evaluations

Not unexpectedly, the press to have definitive outcomes from systemic work and other major social investments led to discussions regarding how complex initiatives were evaluated. *Theories of change* were posited as one new approach. Introduced in the mid-1990s by the Aspen Institute's Roundtable of Comprehensive Community Initiatives for Children and Families, they have served as a viable frame for both planning and evaluating complex efforts. The concept of a theory of change is anchored in the evaluation research advanced by Weiss

(1995), who notes that a theory of change delineates why and how an initiative works. The Aspen group elaborated on this, describing a theory of change as a "systematic and cumulative study of the links between activities, outcomes and contexts of the initiative" (Connell & Kubisch, 1998). Contemporary for its time and still fresh, this approach recommends discerning the desired outcomes of an initiative at the outset, taking into consideration contextual variation.

In the realm of ECEC specifically, Coffman (2012) has advanced a workable theory of change for evaluating systems work. She suggests as her overall theory that systems-building activities will lead to a better early childhood system, which in turn will lead to better child and family outcomes. Contending that there are five elements of system-building (context, components, connection, infrastructure, and scale), Coffman acknowledges that just as there is no one best system, there is no one best way to evaluate systems. She does call for serious accountability efforts, however, suggesting that systemic outcomes will precede comprehensive child and family outcomes.

A Re-clarification of the Need for and Content of Systemic Outputs/Outcomes

Theory of change work has been instrumental in directing those concerned with generating and using research to think far more systematically about outcomes that accrue not only for children and families, but for the system itself. Conventionally, outcomes have been focused on two primary areas: program quality and child performance. Too numerous to chronicle here, legions of instruments exist to help practitioners and policymakers assess the quality of ECEC settings. Similarly, many instruments exist to assess children's development and accomplishments. Instruments that assess children's language and cognitive development have been joined by those that assess children's social and emotional development, and their approaches to learning/executive functioning. Yet as robust has these efforts have been, there has been remarkably little attention devoted to defining and assessing the outputs and outcomes associated with systems work.

To address this void, a distinction between ECEC systemic outputs and outcomes was identified, with the overall outcome defined as child and family well-being. Two sets of outputs contribute to this outcome: family outputs and systemic outputs, which were initially delineated as quality, equitable distribution, and sustainability of services for young children and their families (Kagan, 2015). Originally posited as part of a theory of change, the systemic outputs were expanded to include efficiency in a study of early childhood systems in Latin America (Kagan, Araujo, Jaimovich, & Aguayo, 2016) and later refined for use in a global study (Kagan, 2018). The most recent version of the theory of change begins with the outcomes, suggesting that the ultimate outcome, positive child and family well-being (F), is predicated on a combination of family supports (E) and four systemic outputs (D) (letters correspond to Figure 2.1). These four systemic outputs call for programs and services that are: (i) high-quality; (ii) equitably distributed; (iii) sustainable; and (iv) efficient. These outputs can only be produced when an effective system (C) exists. Such a system is based on a clearly delineated infrastructure (B) that supports diverse programs and services (A), sometimes linked by a boundary-spanning mechanism that integrates programs and services across departmental boundaries. All malleable, these factors are encased in both temporal (political, economic, environmental) (G) and socio-cultural (values, beliefs, heritages, religions) contexts (H). Embedded in this theory of change, then, is a blatant focus on the creation of an ECEC system as the essential prelude for achieving broad-scale outcomes and systemic outputs for children and their families. The theory of change explicitly positions systems work at the heart of efforts to achieve successful outcomes.

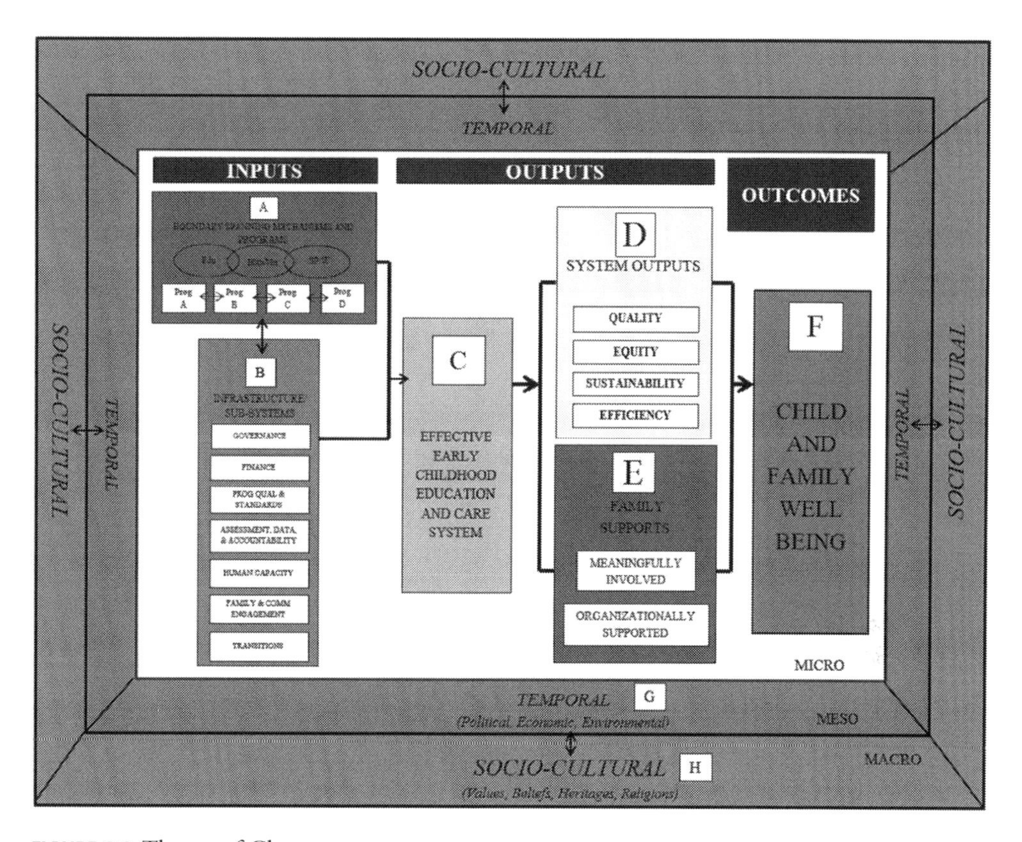

FIGURE 2.1 Theory of Change

As one reviews the evolution of systems thinking in ECEC, several conclusions can be drawn. First, although stemming from practical need, systems work in early childhood has in recent years become far more scientifically and theoretically grounded. It has adopted approaches from sister sciences, and has created new ways to better understand the complex phenomena associated with systemic reform. Second, a nascent field of ECEC systems work is emerging, replete with dissertations on ECEC systems, scholarly publications, and rich debates. Informed by domestic and international systems studies, the scholarship is not restricted by geographic locales, further enhancing the rich contextual tapestry that is being considered. Third, ECEC systems study is pliant, and it is likely to be differently extruded over time and place. It is unlikely to follow a stringent scientific dictum typically associated with a preferred, gold-standard methodology. But drawing from post-modernist views, ECEC systems science, however nascent, is taking shape. It is a science to watch as it is contoured by emerging scholars who understand its value and are sufficiently audacious to undertake its challenges.

Part V – The Effects and Future of Systems Science in ECEC

The Effects of ECEC Systems Work:Serendipity in Action

As the foregoing section suggests, there have been notable advances in ECEC systems thinking, systems work, and systems theory over the past decades. Not all have happened in unison or even in a planned, linear, or even predictable way. Characterized to some extent by happenstance, systems efforts have bumped up against each other, sometimes influencing others haltingly and sometimes with more alacrity. To best understand the rise and effects of a systems orientation in ECEC, it is necessary to provide the gestalt of the change, first by chronicling field-based developments and then by addressing theoretical advances.

ECEC Systems Science and Theory

As this chapter has suggested, systems thinking has had a profound effect on the way ECEC is now conceptualized. No longer content to fund only programs, the public sector and private philanthropy have taken leadership roles in advancing systems approaches. These forward-thinking efforts are loosely adapted from the theoretical framing of systems science, but are innovative in applying it specifically to ECEC. Conceptually, this adaptation acknowledges that ECEC systems are composed of a number of sub-systems (e.g., governance, finance, accountability, professional development, family engagement, regulation, and transitions), each with its own parameters and theories. To render the complexity of systems more transparent and make theoretical work more usable, metaphors and diagrams (e.g., flowers and gears, ovals, theory of change charts) have been developed, a necessary first step that precedes implementation in the field.

Theoretical work of the past three decades has had considerable impact on the way policymakers are conceptualizing services to the young, although the complex policymaking apparatus, replete with topically segregated committee structures and operationalized in striated departments at the federal and state levels, means that constructing policy based on systems theory is difficult to do. Theory-making in ECEC systems work has benefited from work in allied sciences including organizational development, sociology, and psychology; it has also been driven by external forces in the social zeitgeist, including the press for accountability and data and the need for efficiency and sustainability. Whatever the rationale, ECEC systems theory now exists at a broad level and has become sufficiently popularized so that it is quite accessible for practitioners. This has lent both legitimacy and cache to ECEC systems work.

ECEC Systems Science and Practice

Born from need and fueled by both the public and private sectors, ECEC systems work is robust and takes many forms. There is hardly a state that is not engaged in some systems-building effort presently. Indeed, some worry that the efforts are so plentiful, with each taking a unique focus, that their proliferation actually may inhibit the development of a cohesive system. Yet, as Satkowski (2009) notes, "While this proliferation of collaboration initiatives and councils may seem counterproductive to a goal of creating a unified system of early childhood, the efforts can work in harmony, and in many states, they do." There is much reason to be optimistic about the direction of systems work in the United States.

Of course, implementing new ECEC systems work is not only complicated by their number, but by their design and intentions. Sometimes the unit of change is the whole ECEC system, but more frequently the focus of the reform is a single sub-system (e.g., governance, professional development, or accountability). Sometimes the work is externally incentivized (via the ELCF, for example) and sometimes it is quite internally driven. More difficult to advance than service-driven policy (e.g., universal pre-kindergarten), systems work is too often not a policy priority. Indeed, it often remains a hard policy sell, in part due to the ever more apparent implementation challenges that accompany it. ECEC systems efforts are hard to mount; they often take time and demand functional and structural restructuring.

Consequently, there are associated costs with systems instantiation, costs that are often too heavy for constrained government budgets to bear. ECEC systems-building efforts are also hindered by political turnover, as they may go from being a favored approach of one administration to being the Achilles' heel of another. Moreover, ECEC systems efforts are hard to evaluate. Though progress has been made on examining their implementation (Kagan, 2018; Kagan & Landsberg, 2019), there is far less data available on systemic outcomes and limited scholarship being advanced to examine these outcomes as a necessary prelude to child and family outcomes.

The Future of ECEC Systems Work

Advancing Conceptualization of a System

It has become quite fashionable to discuss and study systems work, as the bevy of recent reports cited above indicate. Yet upon more detailed scrutiny, the systems studied in these publications are often conceptualized as governance mechanisms, financing mechanisms, or professional development mechanisms. These strategic approaches denote progress, but they are doing so at the sub-system level, rather than advancing the ECEC system in its entirety. Certainly, such efforts may, as a by-product, lubricate overall systems reform, but they do not substitute for the essential synergy and relationships that are associated with conceptualizing and enacting a total system. Quite prevalent, this sub-systems focus results from operational realities such as funding and time constraints that prevent total systemic reform. While such a sub-system emphasis is therefore realistic, it is also dangerous. It is realistic because total systems work is hard to accomplish; it is dangerous because improvements in any single sub-system cannot alone yield the desired systemic outcomes of quality, equity, sustainability, and efficiency. In short, we need to underscore the importance of broad systemic visions, and emphasize that implementation and analysis of sub-system strategies should not be mistaken for full systems work.

Using the Past to Guide Future ECEC Systems Work

Over three decades, the impetus for ECEC systems work has become solidified, though its acceptance has emerged more from practical need than from scientific inquiry. As such, some might consider ECEC systems work either a-theoretical or lacking a science. Arguably correct, these criticisms may be strengths, however. That the work has emerged from on-the-ground need and practice fosters its legitimacy and potentially its durability. Moreover, ECEC systems science, to the degree that it exists, is the child of many parents; it comes from diverse sciences and fields and has the option of taking the best points from each and adapting them, thus rendering a truly unique approach that fits the political and social history of the field.

Nonetheless, persistent challenges demand consideration. First, lacking a codified scientific base, diverse systemic approaches emerge, each with compelling rationales and objectives; as described above, this is in some ways an advantage. But the fact that ECEC systems work is anchored in different theories also means that time, energy, and funds are lost bringing diverse players and ideologies to consensus. It also means that rarely are truly comprehensive systemic plans developed and implemented. Rather, there is a cacophony of systems efforts. And because systems work is becoming increasingly responsive to context, there is no one model system that fits all settings, nor one "ideal" systemic vision to emulate (Kagan &

Landsberg, 2019). Thus, in being contextually responsive, the field sacrifices the idea of a finely tuned "model system." What it can offer, however, are examples of better developed sub-systems.

Moreover, there is no mega-theory of change guiding implementation. Understanding that all the sub-systems can't be constructed at once, implementers face hard questions – for which there are no empirical answers – regarding which sub-system is best to launch first, and second, and so on. There is no cascade theory that examines how implementation of one sub-system may impact others, and there is no concrete guidance on the fiscal resources necessary for full implementation, although a recent National Academies of Sciences, Engineering, and Medicine (2018) report does suggest a potential method for discerning this.

Quirky, But Essential, ECEC Systems Science

An essential principle reverberating through this chapter is the reality that ECEC systems work is loosely – but only loosely – tied to hard science. Its pulse and energy emanate from the field. As such, ECEC systems work remains somewhat handicapped, if having a codified science is a metric of and for advancement. Nevertheless, this analysis reveals that ECEC systems work has come a long way, moving from near obscurity to understanding, and from chaos and confusion to more nuanced implementation.

Yet, if the systems work is to progress through the twenty-first century, the science of ECEC systems needs critical attention now. At a minimum, several steps are necessary. First, there needs to be some agreement on the desired outcomes of an ECEC system. Individual scholars have proffered such heuristics, but there is no broad-based agreement on them. Second, measurement science must take root in ECEC systems work so that, once agreed-upon measures of outcomes are defined and operationalized, they can be effectively measured. This step needs to be accompanied by the development of comprehensive systems for the collection, analysis, and dissemination of data. Third, evaluation sciences must be marshalled to create codified ECEC systems research. This must be supported through reliable and sustained funding, so that results from different approaches to systems work can be evaluated. The goal of such evaluations should not be merely to discern what works and what doesn't, but to tease out the contextual conditions and process markers that lead to successful implementation and successful outcomes. Scholars need to be trained for this work, and professional associations must accord more attention to promoting scholarship of this sort in their journals and conferences. Analyses of how fields develop (Reid, 2018) suggests that these conditions are requisite for a field to come of age.

ECEC systems work is at the cusp; no longer a toddler struggling to walk straight or a young child delighting in experimentation and awash with the joys of trial and error approaches to learning, ECEC systems work is in its adolescence, changing shape, seeking identity, and developing character. Like adolescence, it offers untold promise, but demands careful support.

References

Atchison, B., & Diffey, L. (2018). *Governance in early childhood education*. Denver, CO: Education Commission of the States. Retrieved from ecs.org/wp-content/uploads/Governance-in-Early-Child hood-Education.pdf

Bellinger, G. (2004). *Systems thinking: The way of systems*. Systems Thinking. Retrieved from system s-thinking.org/systhink/systhink.htm

Banathy, B.H. (1995). Developing a systems view of education. *Educational Technology*, 35(3), 53–57.
Bertalanffy, L. (1972). The history and status of general systems theory. *The Academy of Management Journal*, 15(4), General systems theory, 407–426.
Bipartisan Policy Center. (2018). *Creating an integrated efficient early care and education system to support children and families: A state-by-state analysis.* Washington, DC: Author.
Borgatti, S.P., Mehra, A., Brass, D.J., & Labianca, G. (2009). Network analysis in the social sciences, *Science*, 323(5916), 892–895.
Bruner, C. (n.d.). A framework for state leadership and action in building the components of an early childhood system. Build Initiative. Retrieved from buildinitiative.org/Portals/0/Uploads/Documents/Four_Ovals_paper.pdf
Campbell, F.A., & Ramey, C.T. (1994). Effects of early intervention on intellectual and academic achievement: A follow-up study of children from low-income families. *Child Development*, 65, 684–698.
Coffman, J. (2012). Evaluating systems building efforts. In S.L. Kagan & K. Kauerz (Eds.), *Early childhood systems: Transforming early learning* (pp. 199–215). New York, NY: Teachers College Press.
Connell, J., & Kubisch, A. (1998). Applying a theory of change evaluation approach to the evaluation of comprehensive community initiatives: Program, prospects, and problems. In K. Fulbright-Anderson, A. Kubisch, & J. Connell (Eds.), *New approaches to evaluating community initiatives: Theory, measurement, and analysis* (pp. 15–44). Aspen, CO: The Aspen Institute.
Dichter, H. (2015). Governance as a driver for systems development: Issues of scope and implementation. In S.L. Kagan & R.E. Gomez (Eds.), *Early childhood governance: Choices and consequences* (pp. 58–65). New York, NY: Teachers College Press.
Early Childhood Data Collaborative (2010). *A framework for state policymakers: Building and using coordinated state early care and education data systems.* Available from the Center for the Study of Child Care Employment, Berkeley, CA.
Gallagher, J., Harbin, G., Thomas, D., Clifford, R., & Wenger, M. (1988). *Major policy issues in implementing Part H P.L. 99–457 (infants and toddlers).* Chapel Hill: University of North Carolina Institute for Child and Family Policy.
Gomez, R.E. (2015). Governance as a lever for bringing coherence to ECE systems: The adaptive capacities of consolidated approaches. In S.L. Kagan & R.E. Gomez (Eds.), *Early childhood governance: Choices and consequences* (pp. 45–57). New York, NY: Teachers College Press.
Gruendel, J., & Stedron, J. (2012). Early childhood data systems: a platform for better outcomes for children and programs. In S.L. Kagan & K. Kauerz (Eds.), *Early childhood systems: Transforming early learning* (pp. 119–133). New York, NY: Teachers College Press.
Hayes, C., Palmer, J., & Zaslow, M. (1990), *Who cares for America's children? Child care policy for the 1990s.* Washington, DC: National Academy Press.
International Federation for Systems Research (IFSR). (n.d.). Founding of the IFSR. Retrieved from ifsr.org/what-is-the-ifsr/founding-of-the-ifsr/
Jordan, E., & King, C. (2015). *Rising to the challenge: Building effective systems for young children and families.* The BUILD Initiative. Retrieved from buildinitiative.org/OurWork/StateandLocal/EarlyLearningChallenge.aspx
Kagan, S.L. (2012). Early learning and development standards: An elixir for early childhood systems reform. In S.L. Kagan & K. Kauerz (Eds.), *Early childhood systems: Transforming early learning* (pp. 55–70). New York, NY: Teachers College Press.
Kagan, S.L. (2015). Reconceptualizing ECE governance: Not the elephant in the room. In S.L. Kagan, & R.E. Gomez (Eds.), *Early childhood governance: Choices and consequences* (pp. 9–29). New York, NY: Teachers College Press.
Kagan, S.L. (2018). *The early advantage: Early childhood systems that lead by example.* New York, NY: Teachers College Press.
Kagan, S.L., & Cohen, N.E. (1996). A vision for a quality early care and education system: Quality programs and a quality infrastructure. In S.L. Kagan & N.E. Cohen (Eds.), *Reinventing early care and education* (pp. 309–332). San Francisco, CA: Jossey-Bass.
Kagan, S.L., & Cohen, N.E. (1997). *Not by chance: Creating an early care and education system for America's children – Abridged report.* New Haven, CT: Bush Center in Child Development and Social Policy.

Kagan, S.L., & The Essential Functions and Change Strategies Task Force. (1993). *The essential functions of the early care and education system: Rationale and definition.* New Haven, CT: Yale University Bush Center in Child Development and Social Policy.
Kagan, S.L., & Gomez, R.E. (Eds.). (2015). *Early childhood governance: Choices and consequences.* New York, NY: Teachers College Press.
Kagan, S.L., & Landsberg, E. (2019). *The early advantage 2: Building systems that work for young children.* New York, NY: Teachers College Press.
Kagan, S.L., & Neville, P.R. (1993). *Integrating services for children and families: Understanding the past to shape the future.* New Haven, CT: Yale University Press.
Kagan, S.L., Araujo, M.C., Jaimovich, A., & Aguayo, Y.C. (2016). Understanding systems theory and thinking: Early childhood education in Latin America and the Caribbean. In A. Farrell, S.L., Kagan, & E.K.M. Tisdall (Eds.), *The SAGE handbook of early childhood research* (pp. 163–184). London: SAGE.
Kauerz, K., & Kagan, S.L. (2012). Governance and early childhood systems: Different forma, similar goals. In S.L. Kagan & K. Kauerz (Eds.), *Early childhood systems: Transforming early learning* (pp. 87–103). New York, NY: Teachers College Press.
Klir, G. (1991). *Facets of systems science.* New York: Springer Science+Business Media New York.
Lindblom, C., & Cohen, D. (1979). *Usable knowledge: Social science and social problem solving.* New Haven, CT: Yale University Press.
Mobus, G.E., & Kalton, M.C. (2014). *Principles of systems science.* New York: Springer.
National Academies of Sciences, Engineering, and Medicine. (2018). *Transforming the financing of early care and education.* Washington, D.C.: The National Academies Press.
Neuman, M.J., Roth, J., & Kagan, S.L. (in press). A compendium of international early childhood systems research. Unpublished manuscript, Teachers College, Columbia University.
Regenstein, E. (2015). Glancing at governance: The contemporary landscape. In S.L. Kagan & R.E. Gomez (Eds.), *Early childhood governance: Choices and consequences* (pp. 33–44). New York, NY: Teachers College Press.
Reid, J.L. (2018). Advancing the field of ECP: Lessons from other fields. Unpublished report, Early Childhood Policy in Institutions of Higher Education, Teachers College, Columbia University, New York.
Rousseau, D. (2015). General systems theory: Its present and potential. *Systems Research and Behavioral Science*, 32(5), 522–533.
Satkowski, C. (2009). The next step in systems-building. New America. Retrieved from newamerica.org/education-policy/policy-papers/the-next-step-in-systems-building/
Schaack, D., Tarrant, K., Boller, K., & Tout, K. (2012). Quality rating and improvement systems: Frameworks for early care and education systems reform. In S.L. Kagan & K. Kauerz (Eds.), *Early childhood systems: Transforming early learning* (pp. 71–86). New York, NY: Teachers College Press.
Schweinhart, L.J., Barnes, H.V., Weikart, D.P., Barnett, W.S., & Epstein, A.S. (1993). *Significant benefits: The High/Scope Perry Preschool study through age 27.* Ypsilanti, MI: High/Scope Press.
Sugarman, J. (1991). *Building early childhood systems: A resource handbook.* Washington DC: Child Welfare League of America.
Tout, K., Starr, R., Soli, M., Moodie, S., Kirby, G., & Boller, K. (2010). *Compendium of quality rating systems and evaluations.* Produced for the Office of Planning, Research and Evaluation, Administration for Children and Families, U.S. Department of Health and Human Services.
von Glasersfeld, E. (1981). An introduction to radical constructivism. In P. Watzlawick (Ed.), *The invented reality* (pp. 17–40). English translation. New York: Norton.
Webster (1970). Definition of "System." *Webster's new world dictionary* (2nd college ed.).
Weiss, C.H. (1995). Nothing as practical as good theory: Exploring theory-based evaluation for comprehensive community initiatives for children and families. In J. Connell, A. Kubish, L. Schorr, & C. Weiss (Eds.), *New approaches to evaluating community initiatives* (pp. 65–91). Washington, DC: Aspen Institute.
Yoshikawa, H., Wuermli, A.J., Raikes, A., Kim, S., & Kabay, S.B. (2018). Toward high-quality early childhood development programs and policies at national scale: Directions for research in global contexts. *Social Policy Report*, 31(1).

3

POLICY AND POLITICAL INFLUENCES ON EARLY CHILDHOOD EDUCATION

Jacqueline Jones and Sara Vecchiotti

Ideally, public policy related to early childhood education (ECE) should be driven by research from scientific domains such as developmental psychology, neuroscience, epigenetics, and economics. However, the relationship between science and policy does not always follow a linear trajectory. As public attitudes have changed, so have our expectations for programs for young children and our expectations of their outcomes for children and families. Given the multiple domains that contribute to ECE, different scientific notions can compete simultaneously for attention. Scientific ideas often stew for a while until a window of opportunity permits a connection to policy. On occasion, a policy itself requires that research be conducted to examine that policy's impact or rationale. The relationship of scientific findings to early childhood policy is complicated further in the United States by the lack of a national public policy for early childhood education, related to its purpose, implementation, or outcome.

This chapter will review the historical relationship between science from multiple disciplines such as psychology, neuroscience and epigenetics, and economics, and public policy related to early childhood education. While focused on the scientific knowledge that has played a role in legislation, funding and program implementation, instances in which public policy has influenced new research initiatives will be discussed. Current concerns about new research questions, metrics and methods will be presented.

The Trajectory of Scientific Thinking

Scientific inquiry from a variety of domains have informed and shaped early childhood education and care. They include psychology, neuroscience and epigenetics, and economics has contributed to a public narrative and policy regarding the care and education of young children.

Psychology

The twentieth century witnessed the transition of the United States from an agrarian to an industrial nation, and public policy regarding the care and education of young children was often driven by scientific findings and shifting societal attitudes around child care, parenting, welfare, work, and individual responsibility. As the American population moved from farms to urban areas, public policy was inclined to provide financial support to women who were widowed, abandoned, and without the means to support themselves or their children. The early underlying assumption was that children were better off at home being cared for by their mothers (Blank & Blum, 1997; Jones, 2018; Stevens, 2015). However public attitudes and public policy shifted with the growing numbers of unwed mothers and people of color who began to receive government subsidies. Stevens (2015) reports that:

> By 1995, 10 percent of all American mothers-including 7 percent of white mothers, 20 percent of Hispanic mothers, and 25 percent of African American mothers-were on Aid to Families with Dependent Children (AFDC).[1] Almost half had never been married.
>
> *(p.20)*

The early childhood policy agenda that provided direct payments to support poverty-stricken and widowed women so that they could remain at home and care for their children was challenged by public concern regarding the increasing numbers of Americans who were perceived to be taking advantage of public assistance.

Additional societal changes were occurring. During World War II women entered the paid workforce in large numbers, and the need increased for child care outside of the home (U.S. Bureau of Labor Statistics, 2009). This trend continued into the twenty-first century. Ruhm (2011) reported that, "Sixty percent of mothers with children under the age of six worked in 2008 compared with 33 percent in 1975" (p. 38). Over time, policy leaned towards using public funds to support increasing access to early childhood programs that were conducted outside of the home. The primary intent of many of these programs was to provide a safe environment while parents were working. However, for children whose families were living in poverty, early childhood education programs outside of the home were being perceived as educational interventions that could alter the trajectory of children's development, learning, and success in life.

While many of the early ECE intervention programs targeted specific subgroups of young children, such as African American children whose families were living in poverty, the policy discussion moved to a debate on the usefulness of providing universal access within a city, state, or group of school districts. Why ECE programs should receive public funds and the expected outcomes for young children and their families continues to change over time.

Intelligence: From Fixed to Malleable

During the first half of the twentieth century, the poor academic performance of children living in poverty compared with that of their more resourced peers was attributed to innate cognitive deficits. From a scientific perspective, intelligence was fixed, and biology determined development. Some scientists linked intelligence to ethnic and racial background (Jensen, 1969). However, there was an emerging perspective that intelligence may be malleable, and researchers were investigating the impact of targeted intervention programs on the intellectual development of young children whose families were living in poverty (C. Deutsch, 1965; M. Deutsch, 1964; Gray & Klaus, 1965; Hunt, 1961).

1 Aid to Families with Dependent Children (AFDC) was a federal assistance program in effect from 1935 to 1996 created by the Social Security Act (SSA) and administered by the United States Department of Health and Human Services that provided financial assistance to children whose families had low or no income.

The 1960s marked the emergence of some of the seminal model early childhood intervention programs as well as implementation of some of ECE's landmark federal legislation, funding, and program implementation. The ground-breaking Perry preschool program (W. Steven Barnett, 1996) and the North Carolina Abecedarian program (Ramey & Campbell, 1984; Ramey et al., 2000) were specifically designed to provide empirical evidence on the short- and long-term effects of intervention programs on African American children who were living in conditions of poverty. In 1967 the Chicago-Public School District became the first district to use Title I funds under the Elementary and Secondary Education Act to establish the Child–Parent Centers (CPC), a model early childhood preschool program, which demonstrated strong long-term outcomes for young children (Reynolds, 2000; Reynolds, Ou, & Temple, 2018; Reynolds & Wolfe, 1997). The early childhood research base of the early twenty-first century has its roots in the longitudinal results of the Perry Preschool and Abecedarian projects, and the Chicago Child–Parent Centers. These three remain the gold standard of early childhood education, and they laid the foundation for the federally funded Head Start program, which began in 1965, and the Elementary and Secondary Education Act of 1965. While these early programs were initially targeted to support poor children and children of color, over time this trend shifted to discussions of whether these programs might benefit all children, including more financially well-off and middle-class families who were increasingly unable to afford the rising cost of child care.

Neuroscience and Epigenetics: The Developing Brain Needs Nurturing Relationships and Positive Experiences

The 2000 National Academies publication of *Neurons to Neighborhoods* (National Research Council and Institute of Medicine, 2000) placed a significant focus on neuroscience as a source of information on understanding early development and advocating for increased funding for early childhood education. The report outlined ten core concepts or principles that guide the developmental process:

1. Human development is shaped by a dynamic and continuous interaction between biology and environment.
2. Culture influences every aspect of human development and is reflected in childrearing beliefs and practices designed to promote healthy adaptation.
3. The growth of self-regulation is a cornerstone of early childhood development that cuts across all domains of behavior.
4. Children are active participants in their own development, reflecting the intrinsic human drive to explore and master one's environment.
5. Human relationships, and the effects of relationships on relationships, are the building blocks of healthy development.
6. The broad range of individual differences among young children often makes it difficult to distinguish normal variations from maturational delays from transient disorders and persistent impairments.
7. The development of children unfolds along individual pathways whose trajectories are characterized by continuities and discontinuities, as well as by a series of significant transitions.
8. Human development is shaped by the ongoing interplay among sources of vulnerability and sources of resilience.

9. The timing of early experiences can matter, but, more often than not, the developing child remains vulnerable to risks and open to protective influences throughout the early years of life and into adulthood.
10. The course of development can be altered in early childhood by effective interventions that change the balance between risk and protection, thereby shifting the odds in favor of more adaptive outcomes.

These concepts strongly suggested that the right intervention at the right time in a child's development could improve opportunity for success in school and in life. Strategic dissemination and clear messaging of the report's recommendations to policy makers and practitioners resulted in widespread attention to the report's findings.

In 2015 the Institute of Medicine released a consensus report (Institute of Medicine and National Research Council, 2015) that was designed to outline a vision for unifying the early childhood workforce. Beginning with a review of the science of early childhood from the 2000 release of *Neurons to Neighborhoods* to 2015, the report outlined a set of competencies that teachers of young children should master in order to support healthy development and learning in young children. These competencies were based on four broad categories of insight that were outlined in the 2015 report and that reflected the current state of the science of early childhood education:

> First, the past decade of research has converged on an understanding that in many or perhaps even most instances, causality with respect to disease, disorders, and maladaptive development – as well as the preservation of health and maintenance of normative, adaptive development – is best viewed as an *interplay between genome-based biology and environmental exposures*. This understanding represents a clear departure from the historical views that human morbidities are attributable to either pathogenic environments or faulty genes...
>
> Second, the role of *developmental time* in the dramatic unfolding of brain structure and function and the acquisition of concomitant human capacities has become increasingly important in explaining early development. Critical and sensitive periods – time windows in which experience-related developmental transitions must or can most readily occur – create a temporal mapping of anticipated early childhood exposures that guide the timing and sequencing of developmental change. As the molecular substrates for such critical periods and events have become known and tractable, altering their timing and manipulating the opening and closing of specific developmental windows have become increasingly plausible (Greenough and Black, 1992; Greenough et al., 1987; NRC and IOM, 2000).
>
> Third, there is now strong evidence that *early psychosocial adversities* – beginning even during fetal development – can have important short- and long-term effects on the brain's development, the regulation of stress responsive hormone systems, and the calibration of stress reactivity at a variety of levels from behavioral to gene expression responses...
>
> Fourth, inquiry into the sources of special vulnerability and resilience with respect to early adversity has led to the discovery of substantial *individual differences in children's susceptibilities to both negative and positive environmental exposures.*
>
> *(Institute of Medicine and National Research Council, 2015, pp. 58–59; emphasis in original)*

Economics: Investments in ECE Can Lead to Significant Returns

While psychology and neuroscience proposed ECE as an intervention with the possible ability to boost intellectual potential and mitigate the effects of living in poverty, economics provided a more practical perspective. Perhaps the most widely cited work is based on the research of Nobel Prize winning economist, James Heckman. Reviewing data from several decades of research on the Perry Preschool Project, Heckman concluded that the short- and long-term benefits of high-quality ECE such as Perry, were strong enough to think of ECE as an *investment* in young children's learning and development that would yield a significant *return*, not just to individual children, but to society at large (Heckman, Moon, Pinto, Savelyev, & Yavitz, 2010). Economics introduced the language of return on investment (ROI) and cost/benefit analysis into the ECE discourse (Barnett & Nores, 2015). Even a Presidential State of the Union Address (Obama, 2013) used economic return as the basis for proposing enhanced ECE policies:

> I propose working with states to make high-quality preschool available to every child in America. Every dollar we invest in high-quality early education can save more than seven dollars later on – by boosting graduation rates, reducing teen pregnancy, even reducing violent crime.

However, questions remain, as: "this evidence does not quash debates regarding the degree to which these services should be universal or targeted, the amount of the return on investments (ROI), and the generalizability of these research findings" (Kagan, Gomez, & Roth, 2018, p.3).

Influence of Science on Public Policy Related to the Early Childhood Education

The relationship between science and public policy related to early childhood education is not always direct. Each program or policy has a unique story that is grounded in varying contexts. In most cases there are windows of opportunity that allow scientific inquiry, societal attitudes, and individual champions to play an influential role. Aspects of the relationship between research and policy in ECE may be captured in the stories of three major ECE programs: (1) Head Start, (2) the Child Care Development Fund, and (3) state-funded preschool.

Head Start: Inspired by Research; The Catalyst for a New Research Base

As mentioned previously, in the 1960s the United States was also growing increasingly aware of widespread poverty,[2] especially among young children, and its negative implications for early development and learning. When President Lyndon Johnson's administration waged a War on Poverty in the 1960s, the wellbeing of young children took a prominent position.

2 In 1959, the national poverty rate was over 22 percent; in 1964 it was 19 percent; and by 1974, after President Johnson's War on Poverty, it dropped to 11.2 percent. In 1959, the poverty rate for African Americans was 55.1 percent and by 1969 it dropped to 32.2 percent (https://www.washingtonpost.com/news/wonk/wp/2012/07/11/poverty-in-the-50-years-since-the-other-america-in-five-charts/?utm_term=.b57f120aefb0). In 1959, the child poverty rate was 27.3 percent; in 1964 it was 23 percent; and by 1969 it was 14 percent (http://www.pewresearch.org/fact-tank/2014/01/13/whos-poor-in-america-50-years-into-the-war-on-poverty-a-data-portrait/).

With studies supporting the hypothesis that intelligence was not fixed and the right interventions could ameliorate the impact of child poverty, the US federal government instituted the Head Start program in 1965 within the then Office of Economic Opportunity. The Head Start program included comprehensive health, nutrition, education, and parent involvement services to children and their families. President Johnson stated that Head Start "set out to make certain that poverty's children would not be forevermore poverty's captives...... children are inheritors of poverty's course and not its creators. Unless we act these children will pass it on to the next generation."[3] Although Head Start was initiated as a 6-week summer program to prepare children for kindergarten, it soon become clear that overcoming the early impacts of poverty would require significantly more time. As the program progressed, Head Start grantees operated for a half-day for 10–11 months a year and with new regulations programs are transitioning to full-day, full-year.[4]

From Head Start's inception, communities and neighbourhoods were seen as partners in the effort to mitigate the impacts of poverty.[5] Now under the US Department of Health and Human Services, Head Start funds still flow directly from the federal government to public or private community-based agencies. With over 1,700 public and private agency grantees, the program's budget has grown significantly, and the Consolidated Appropriations Act of 2016 provided $9.17 billion for Head Start in the fiscal year 2016. Through 1,700 grantees, the Head Start and Early Head Start programs provide services to over one million children a year.[6]

Research on Head Start: Is the Program Effective?

From the early phases of the Head Start program, research was commissioned to answer the overarching question of whether participation in Head Start resulted in significantly improved child outcomes. The 1969 Westinghouse Study found marginal effects on children's cognitive development and questioned the effectiveness of the Head Start program. Concerns with the study design led to a re-analysis of the data by Smith and Bissell (1970). They found that children who needed the program the most demonstrated the largest gains. Subsequent research has since determined that the 1969 study was prematurely done, poorly designed, lacked statistical sophistication, and, with hindsight, was too narrowly focused on cognitive outcomes (Campbell & Erlebacher, 1970; Zigler, Styfco, & Gilman, 1993; Zigler & Valentine, 1979). Such investigations continued into the 1980s and found both short- and long-term effects of Head Start not only on children's cognitive development but also on children's health and social development (Cole & Washington, 1986; Darlington, Royce, Snipper, Murray, & Lazar, 1980; Lee, Brooks-Gunn, & Schnur, 1988; McKey, 1985; Wu & Campbell, 1996; Zigler & Berman, 1983).

Yet, findings also suggested that the effects of Head Start were not entirely sustained in the long term (Lazar & Darlington, 1982; Lee, Brooks-Gunn, Schnur, & Liaw, 1990; McKey, 1985). In the 1990s results suggested that the so-called fade-out effect for Head Start might be due to quality differences in subsequent elementary schooling or the home environment (Entwisle, 1995; Lee et al., 1990). Therefore, while research continued to examine whether participation in Head Start programs produced sustained effects, new research questions were linking children's progress to the quality of both Head Start and early elementary classrooms.

3 https://www.presidency.ucsb.edu/documents/remarks-project-head-start
4 The 2016 Head Start Program Performance standards require that all Head Start center-based programs offer at least 1,020 annual hours of service for preschoolers by August 1, 2021, with at least 50 percent of such preschool slots meeting this requirement by August 1, 2019.
5 https://www.presidency.ucsb.edu/documents/remarks-project-head-start
6 https://www.acf.hhs.gov/ohs/about/head-start

Policymakers Call for Program Accountability

Although interest in the long-term impact of Head Start remained strong, there was a growing focus on whether Head Start promoted children's school readiness skills prior to school entry. Policymakers set the research agenda, as seen in 1998, when Congress authorized the Head Start Impact Study to determine whether participation in Head Start programs improved children's school readiness skills in the immediate *and* long-term. In 2005, the US Department of Health and Human Services (HHS) released the first-year results of a large-scale study of the impact of Head Start programs across the country. The results were mixed, with some modest gains demonstrated in early literacy and math skills. The comparison between children enrolled in Head Start programs and children who were not was complicated when researchers reported that over 60 percent of the control group participated in some type of ECE program, including Head Start itself. HHS released the final Head Start Impact Study finding in 2010. These data suggested that the immediate gains in child outcomes had lessened or disappeared by the end of first grade.

In response, stronger accountability regulations were established in 2011. The Designation Renewal System was created in which grant renewal was related to grantee program performance on measures identified by the Department of Health and Human Services. In addition, new regulations were released in 2016 which revised the Head Start Performance Standards to provide "clarity" on how to best support high-quality program delivery. They were developed by "incorporating the growing body of research on effective early care and education."[7]

While large-scale research studies have reported mixed results on the impact of participation in Head Start programs nationwide, some individual programs have demonstrated more positive results. For example, research on the mature Head Start program operated by the Community Action Project of Tulsa, Oklahoma has provided evidence of positive program participation impacts through middle school, primarily through 8th grade for math, grade retention, and absenteeism (Phillips, Gormley, & Anderson, 2016).

Head Start Policy Fuels More Research: The National Reporting System

In 1997, Congress asked the U.S. Government Accountability to evaluate how the Head Start Bureau (HSB) held grantees accountable for performance in achieving program purposes and compliance with laws and regulations (General Accounting Office, 1998). The GAO reported that HSB lacked objective information on the performance of Head Start grantees and whether programs were achieving the desired outcomes for children and families. The 1998 reauthorization of Head Start required the establishment of education standards for which programs could develop local assessments to measure their progress towards achieving. In response, HSB developed the Head Start Outcomes Framework, outlining expectations for children's skills across the developmental domains, to guide programs on-going assessment of children's progress (General Accounting Office, 1998).

In 2002, stemming from the 1998 legislation requiring programs to use the same assessment to report child outcomes twice a year, President Bush's Good Start Grow Smart Initiative announced the creation of a National Reporting System (NRS). The NRS was developed to measure children's progress according to indicators of learning mandated by Congress in 1998 (Government Accountability Office, 2005). During 2002 federal

7 https://eclkc.ohs.acf.hhs.gov/policy/pi/acf-pi-hs-16-04

contractors developed the NRS. A Technical Workgroup of expert advisors was convened to advise the Office of Head Start, and the NRS was implemented in the Fall of 2003 (Government Accountability Office, 2005; Paulsell et al., 2006). The assessment was a 15–20-minute direct child assessment battery focused on vocabulary, letter recognition, simple math skills, and screens to assess children's understanding of English (Vogel, Nogales, Aikens, & Tarullo, 2008).

However, concerns were raised from practitioners, the Head Start community, and members of the Technical Work Group regarding the development of the NRS, its lack of psychometric integrity, inappropriate developmental content, racial, cultural, and linguistic insensitivity, and the high-stakes nature of its intent (Government Accountability Office, 2005; Lazarin & McAdam, 2003; Meisels & Atkins-Burnett, 2004; Moore & Yzaguirre, 2004). Previously, the National Education Goals Panel had recommended that children below the age of eight not be assessed for accountability purposes (Shepard, Kagan, & Wurtz, 1998). Yet, the controversy spotlighted an important research question for the early care and education field as to whether such standardized testing, driven by policy to assess children's progress for program accountability purposes, was appropriate for 4- and 5-year-olds.

Further, the debate drove the ECE research field to ask broad questions about the assessment of young children, such as what were the appropriate purposes for assessment, what was appropriate to measure, what were the appropriate measures, and what methods were appropriate to assess young children's developmental progress (National Research Council, 2008). In 2005, the GAO issued a report citing the development and implementation issues with the NRS and concluded that analysis of the NRS was "incomplete" to support its use for accountability purposes or to target training and technical assistance (Government Accountability Office, 2005, p. 4); they suggested further development was needed.

In 2006 the pushback regarding the NRS and child assessment concerns led congress and the Office of Head Start to request that the National Research Council convene a committee of experts to review research on developmental outcomes and appropriate assessment of young children. The resulting consensus report, *Early Childhood Assessment: Why, What, and How,* (National Research Council, 2008) provided guidelines on the assessment of young children. It has become the definitive guide for determining when and how to assess young children and has been widely cited in legislation related to publicly funded programs that engage in the early childhood assessment.

In the 2007 reauthorization of Head Start, Congress suspended the use of the NRS and instead Congress required the Office of Head Start, as part of their monitoring and review process, to include a valid and reliable observational instrument to assess classroom quality, especially teacher–child interactions. The requirement foreshadowed the creation, the Designation Renewal System, which set a five-year grant period for grantees and included a review to determine whether services provided are high-quality and comprehensive: if not, a grant is subject to award recompetition.[8] Instead of the NRS, the quality of Head Start programs are monitored using the Classroom Assessment Scoring System (CLASS), and poor performance on this measure can force grantees to enter into a recompetition to maintain their grant award.

8 https://eclkc.ohs.acf.hhs.gov/designation-renewal-system

Child Care: Influence of Research Grew Over Time

Child care has a longer history than Head Start, and child care policies sprung from economic necessity, to be followed by intervention research. The first federal investment in child care was in 1933 within the Federal Emergency Relief Administration as a response to poverty during the Great Depression (Cohen, 1996). Primarily, nursery school programs provided government-paid jobs for teachers, cooks, janitors etc. who were unemployed and, as unemployment fell, many of the centers closed (Cohen, 1996). The 1940 Lanham Act funded child care centers to provide affordable care for children so women could work in defense-related industries in World War II (Cohen, 1996). After the war, the centers closed as men returned home to work and women returned home.

It was not until the Comprehensive Child Development Act of 1971 when Congress passed a bipartisan bill to institute a universal child care system modeled after Head Start, that the emerging Head Start child outcome findings influenced child care policy. The purpose of the Act was "to provide every child with a fair and full opportunity to reach his full potential by establishing and expanding child development programs ... to recognize and build upon the experience and success gained through Head Start and similar efforts" (Senate-1512, 1971). This initial research was ignored when President Nixon vetoed the bill.[9] However, the failure to establish a universal child care system set the stage for child care to be continued to be tied to the notion of work supports within public welfare assistance.

In the 1990s, Congress reformed public assistance through the Personal Responsibility and Work Opportunity Reconciliation Act (PWORA) of 1996 and created the Temporary Assistance to Needy Families (TANF) program, emphasizing work requirements tied to time-limited assistance (Boushey, 2001; Long & Clark, 1997; Office of the Assistant Secretary for Planning and Evaluation, 1996). Further, the Child Care and Development Fund (CCDF) was established with mandatary, state-matching funds, and discretionary funds (through the Child Care and Development Block grant) (CCDBG) to provide child cares subsidies (Long & Clark, 1997). Such reforms stemmed from concerns about increasing caseloads and the growing number of women participating in the workforce. Reforms were also based on the 1980s demonstration welfare programs that required training and employment to receive welfare benefits, thereby creating the need to support child care as mothers receiving public assistance began to enter the labor market (Blank & Blum, 1997). Because PWORA and CCDBG have their origins in economic necessity, they laid the foundation for the still persistent tensions between child care being perceived as a work support versus child care intended to focus on learning and development.

Research Influence: Does Child Care Quality Explain Variation in Child Outcomes?

Research on Head Start and other ECE programs in the 1970s, 1980s, and 1990s increasingly led to questions regarding reasons for the variations in child outcomes present in the results (Barnett, 1995; Frede, 1995; Yoshikawa, 1995). For child care, the differences in both

9 Various reasons have been suggested for the veto: 1) internal power plays in White House and Administration staff over control of domestic policy, 2) protection for the President from conservative rivals who cast the bill as Sovietizing American children, 3) deflecting attention from the President's foreign policy agenda in China and the Soviet Union, and 4) appeasing conservatives viewing such a system as weakening or attacking the American family (Roth, 1976).

structural and process quality across programs was thought to be a factor in the variability in child outcomes (Kontos, Hsu, & Dunn, 1994; Kontos, 1991; McCartney, Scarr, Phillips, & Grajek, 1985; Phillips, McCartney, & Scarr, 1987). Research framed structural quality as infrastructure elements such as staff: child ratios, group size, and teacher qualifications, components that form the basis for state and federal child care licensing regulations. Process quality referred to elements the child might experience, such as teacher–child interactions, curricula activities, and peer interactions.

Seminal child care studies examined the relationship between cost, child care quality, and child outcomes (Helburn et al., 1995; Peisner-Feinberg et al., 2000). Reminiscent of the purpose of the Comprehensive Child Development Act, these reports asserted that the child care system should be considered as an "educational program" focused on child well-being, which is in contrast to the view of child care as a work support in PWORA. The 1995 report found that child care quality varied within and between states and that overall the quality of child care was mediocre, which undermined the potential of care to promote children's development. The 1999 follow-up found that children attending higher-quality centers performed better on cognitive and social skill measures through kindergarten and 2nd grade (Peisner-Feinberg et al., 2000). Finally, the report found that classroom practices were related to children's cognitive development, and the quality of teacher–child interactions was related to children's social development.

Most importantly, these studies supported the emerging understanding that high-quality programming is essential to producing the desired child outcomes and that not all ECE programs, especially child care, are of quality. They helped establish a concern for program quality and child outcomes in child care with policymakers. Today, the concern is reflected in policymakers' consistent attention to setting structural quality elements in both child care licensing regulations and workforce competency and qualifications requirements. Quality concerns are still relevant today and many were at least partially responded to until the 2014 reauthorization of the Child Care and Development Block Grant, strengthening health and safety regulations and increasing professional development supports for the ECE workforce (Matthews Schulman, Vogtman, Johnson-Staub, & Blank, 2015).

State-funded Preschool Programs

Preschool Rationale and Research on Long-Term Outcomes

While the federal government was the primary funder of early education, states also began to make significant investments in preschool programs. Starting in the 1960s and 1970s, states began to either expand Head Start programs or create state-funded preschool programs (Mitchell, 2001). In the 1980s, as a response to President Reagan's "A Nation at Risk,"[10] report, which sought educational excellence through public education reform and was based on long-term study results of Perry and Abecedarian (Mitchell, 2001), another 23 states began prekindergarten programs (Mitchell, 2001). By the 1990s, with the focus on school readiness from the National Educational Goals Panel and influenced by emerging neuroscience research indicating the importance of early development and intervention (Bornstein, 1989; Immelmann & Soumi, 1982; National Research Council and Institute of Medicine, 2000),

10 In 1983 President Reagan's National Commission on Excellence in Education released "A Nation at Risk: The Imperative for Educational Reform," reporting that American schools were failing and students were underperforming in academic achievement as compared with students internationally.

up to 42 states established state prekindergarten programs (Mitchell, 2001). The majority of these programs were half-day/part-year programs, were targeted for children from low-income families, and were not of sufficient scale to serve all eligible children (Schulman, Blank, & Ewen, 1999). Today, over 1.5 million children are enrolled in state preschool programs with states spending over $8 billion – more than doubling the investment levels in 2002 (Friedman-Krauss et al., 2019).

Research has had a direct influence on state policy, and the increase in state preschool aligns with the increased research on the short-term and the long-term of model ECE programs. In 1997, New York State's executive summary of the originating universal prekindergarten legislation stated:

> Every child is born with a natural desire to learn .... For some of our children, this innate desire to learn has lost its edge even before they enter the schoolhouse door. While countless studies have documented the lifetime benefit that is provided by early education, too many of New York's children fail to receive the early education experience they need to begin school with the social and cognitive skills necessary for a successful journey through our educational system.
>
> *(Silver & Sanders, 1997)*

The work was based on an understanding of both the child development and research on the long-term child outcomes.

Several lawsuits, which began in the 1990s to challenge state education financing systems and whether states were ensuring that all students received "sound basic education" or "thorough or sufficient education" as required in state constitutions, were also influenced by research on child outcomes (Boylan, 2007; Rebell, Wolff, Kolben, & Holcomb, 2017). Advocates began to discuss preschool as part of a public education system, as a remedy for public education failures, and as a method for reducing achievement gaps of low-income children and educational inequity. Cases such as *Abbott vs. Burke* in New Jersey, *Hoke County Board of Education vs. State* (Leandro) in North Carolina, *Abbeville County School District vs. State* in South Carolina, *Lakeview vs. Huckabee* in Arkansas, *Hancock vs. Driscoll* in Massachusetts (Boylan, 2007; Rebell et al., 2017) brought research from developmental psychology and neuroscience into the legal arena. Many of the lawsuits heard expert testimony discussing and summarizing the long-term child outcomes research from the Perry and Abecedarian programs, and judges have issued findings supporting ECE as a strategy to promote children's development, school success, and life outcomes, particularly for children from low-income families (Boylan, 2007; Rebell et al., 2017).

While the litigation efforts continue for many cases, such as New York, Alaska, Colorado, and Connecticut, to date only *Abbott vs. Burke* resulted in a 1998 court-mandate that established a high-quality ECE program for 3- and 4-year-olds in 31 low-income school districts in New Jersey. Research documenting outcomes and program effectiveness of the Abbott program followed (Barnett, Kwanghee, Youn, & Frede, 2013; Frede, Kwanghee, Barnett, & Figueras, 2009) and continues to provide evidence for continued investment as well as being key to expert testimony in judicial deliberations and decisions.

There is also evidence suggesting that the cost–benefit analysis of programs, particularly Perry Preschool and Abecedarian, which indicates significant return on investments, has spurred increasing federal and state investment in early childhood programs. Barnett (1993; 1996) studied the costs/benefits for the Perry program when students were age 27–28 and found net benefits to the individual students, the general public, and to society as a whole.

RAND also found that the benefits of participation in the Perry Preschool project outweighed program costs in the long-term and generated government savings and societal benefits (Karoly et al., 1998). Simultaneously, there were calls for increased investment recognizing that current funding streams were not supporting the true cost of high-quality care (Gomby et al., 1996). Such considerations are continuing with James Heckman's return on investment/human capital work (Garcia, Heckman, Ermini Leaf, & Prados, 2017; Heckman et al., 2010; Sneha, Garcia, Heckman, & Hojman, 2015) and with the 2018 National Academies report on early care and education financing (National Academies of Sciences, Engineering, Medicine, 2018).

Influence of Research: Child Outcomes and Program Quality

With the increase of state prekindergarten programs, a new emerging research area began to examine several large-scale, public ECE programs brought to scale outside of Head Start and child care. Today, Tulsa and Boston, like Abbot, serve as prime examples demonstrating that high-quality preschool which improves children's development (Yoshikawa et al., 2013) can be brought to scale. Children in Tulsa's prekindergarten program demonstrate moderate to large gains in school readiness skills, especially for Latinx children (Gormley Jr, Gayer, Phillips, & Dawson, 2005; Gormley Jr et al., 2011), while children in Boston's Public School program had improved language, literacy, math, and executive functioning skills (Weiland & Yoshikawa, 2013). Further, several early care and education research-to-practice partnerships have also been created in Boston, Chicago, New York City, Virginia, Louisiana, and Connecticut, with policymakers and researchers collaborating on answering research questions regarding continuous quality program improvement during the scale-up of services provided to young children. Therefore, ECE child outcomes research has influenced state programming and investment decisions to provide state preschool for at least some children.

Implementation research is now being used to examine program implementation in practices and to identify what works (or not), for whom, and under what conditions. The question is no longer whether ECE is effective in general. Rather, the lingering question is how is ECE effective, or not, for diverse populations of children? These *scale-up* questions are also relevant to policymakers, at the local, state, and federal level, as they administer the high-quality programs, strive for continuous program quality improvement, and aim to deliver services that result in positive child outcomes for all children (Barnett, 2017; Chaudry, 2017; Phillips et al., 2017).

Conclusion

Early childhood education can trace its origins to experimental developmental science on children's cognition and intelligence. Neuroscience and epigenetics, and economics, have also made significant contributions to our understanding of the physiological mechanisms of development and the appropriateness of our expectations of the impact of early childhood programs. Head Start, child care and state-funded preschool exist as a result of the convergence of changing societal norms, large federal initiatives, and compelling research from various scientific domains. Ongoing research related to such programs as Head Start and state-funded preschool has also informed what we know about the elements of effective programs and the types of gains that various subgroups of children can make.

Research had an influence in providing initial evidence for the notion that intelligence is not fixed and that life outcomes are not pre-determined. After programs and policies were in place, research continued to have an influence by documenting benefits, or lack thereof, for children and families and attempting to explain why outcomes were achieved, or not. However, scientific findings, by themselves, do not create or change policy; there must be windows of opportunity that support new legislation, funding and program development.

The programs and policies built from the first experimental science have led to the field of applied developmental science. Researchers are attempting to generate empirical studies that have relevance to policymakers in the early stages of ECE program development and at scale-up. Policymakers and researchers are working toward a more shared vision of the questions that need to be posed as the field continues to struggle with how to best establish, expand, and improve ECE services for all children or groups of children. Over time issues over early care and education have remained the same, as both researchers and policymakers seek answers to questions such as:

- How can programs be more effective?
- What are the short- and long-term outcomes for children? And subgroups of children?
- What programs and/or system characteristics or components lead to benefits for children?
- What classroom practices and instructional strategies and content lead to benefits for children?
- How can programs be implemented at scale and maintain the expected benefits?
- How can ECE systems and services be best coordinated to serve children and families across the developmental continuum? (Gomby, Larner, Stevenson, Lewit, & Behrman, 1995; Phillips et al., 2017; Ramey & Ramey, 1998; Reynolds, Mann, Miedel, & Smokowski, 1997).

Even the significant growth in ECE programs has not resulted in the degree of impact for young children that was anticipated. Edward Zigler cautioned that "Head Start cannot single-handedly fix broken families, raise incomes, quell neighborhood violence, improve lifetime health care and nutrition, and provide the multitude of enriching experiences that middle-class children have before they set foot in preschool" (Zigler, 2011, p.198). The Center on the Developing Child at Harvard University (2016, p.6) argued that "the average magnitude of intervention effects has not increased substantially in 50 years, while the challenges most current programs were originally designed to address have become even more complex." They explain that "disparities in achievement have grown wider not because early childhood programs have had no impact, but because the size of their effects has failed to keep pace with the benefits of growing up in a high-income family in a rapidly changing world."

Acknowledging the complexity of need, the proliferation of ECE programs, the lack of coordinated service systems, and, as more programs are brought to scale and more children are served, the ability to achieve greater impacts for young children rests on a more nuanced understanding of how programs and program components are being implemented and the differential impacts on subgroups of children. The goal of the next generation of research questions for ECE programs, of interest to both researchers and policymakers, is to explore more intentionally the basic question of the social sciences, "What works, for whom, and under what conditions?"

References

Barnett, W.S. (1993). Benefit-cost analysis of preschool education: Findings from a 25 Year follow-up. *American Journal of Orthopsychiatry*, 63(4), 500–508.

Barnett, W.S. (1995). Long-term effects of early childhood programs on cognitive and school outcomes. *The Future of Children*, 5(3), 25–50. doi:10.2307/1602366

Barnett, W.S. (1996). Lives in the balance: Age-27 benefit-cost analysis of the High/Scope Perry Preschool Program. *Monographs of the High/Scope Educational Research Foundation*, 11, 1–105.

Barnett, W.S. (2017). Challenges to scaling up effective pre-kindergarten programs. In *The current state of scientific knowledge on pre-kindergarten effects* (pp. 67–74). Washington, DC: Brookings Institution and Duke Center for Child and Family Policy.

Barnett, W.S., & Nores, M. (2015). Investment and productivity arguments for ECE. In T.M. Marope & Y. Kaga (Eds.), *Investing against evidence: The global state of early childhood care and education* (pp. 71–86). Paris: United Nations Educational, Scientific and Cultural Organization.

Barnett, W.S., Kwanghee, J., Youn, M.J., & Frede, E. (2013). Abbott Preschool Program Longitudinal Effects Study: Fifth grade follow-up. Retrieved from National Institute for Early Education Research: http://nieer.org/research-report/201311apples205th20grade-pdf

Blank, S.W., & Blum, B.B. (1997). A brief history of work expectations for welfare mothers. *The Future of Children*, 7(1), 28–38.

Bornstein, M.H. (1989). Sensitive periods in development: Structural characteristics and causal interpretations. *Psychological Bulletin*, 105(2), 179–197.

Boushey, H. (2001). *The effects of the personal responsibility and work opportunity reconciliation act on working families: Viewpoints*. Washington, DC: Economic Policy Institute. Retrieved from https://www.epi.org/publication/webfeatures_viewpoints_tanf_testimony/

Boylan, E. (2007). High-quality prekindergarten as the first step in educational adequacy: Using the courts to expand access to state preK programs. *Children's Legal Rights Journal*, 27(1), 24–55.

Campbell, D., & Erlebacher, A. (1970). How regression artifacts in quasi-experimental evaluations can mistakenly make compensatory education look harmful. In J. Hellmuth (Ed.), *Compensatory education: A national debate* (Vol. 3). New York: Brunner Mazel.

Center on the Developing Child at Harvard University. (2016). From best practices to breakthrough impacts: A science-based approach to building a more promising future for young children and families. Retrieved from http://www.developingchild.harvard.edu: http://www.developingchild.harvard.edu

Chaudry, A. (2017). The promise of preschool education: Challenges for policy and governance. In *The current state of scientific knowledge on pre-kindergarten effects*. Washington, DC: Brookings Institution and Duke Center for Child and Family Policy.

Cohen, A.J. (1996). A brief history of federal financing for child care in the United States. *The Future of Children*, 6(2), 26–40. doi:10.2307/1602417

Cole, O.J., & Washington, V. (1986). A critical analysis of the assessment of the effects of Head Start on minority children. *Journal of Negro Education*, 55(1), 91–106. doi:10.2307/2294636

Darlington, R.B., Royce, J.M., Snipper, A.S., Murray, H.W., & Lazar, I. (1980). Preschool programs and later school competence of children from low-income families. *Science*, 208(4440), 202–204. doi:10.1126/science.208.4440.202

Deutsch, C. (1965). Education for disadvantaged groups. *Review of Educational Research*, 35(2), 140–146.

Deutsch, M. (1964). Facilitating development in the pre-school child: Social and psychological perspectives. *Merrill-Palmer Quarterly of Behavior and Development*, 10(3), 249–263.

Entwisle, D.R. (1995). The role of schools in sustaining Early Childhood Program benefits. *The Future of Children*, 5(3), 133–144. doi:10.2307/1602371

Frede, E., Kwanghee, J., Barnett, S.W., & Figueras, A. (2009). The APPLES Blossom: Abbott Preschool Program longitudinal effects study, preliminary results through 2nd grade. Retrieved from National Institute for Early Education Research: http://nieer.org/wp-content/uploads/2012/04/apples_second_grade_results.pdf

Frede, E.C. (1995). The role of program quality in producing early childhood program benefits. *The Future of Children*, 5(3), 115–132. doi:10.2307/1602370

Friedman-Krauss, A.H., Barnett, S.W., Garver, K., Hodges, K., Weisenfeld, G.G., & DiCrecchio, N. (2019). *The state of preschool 2018: State preschool yearbook*. Retrieved from National Institute for Early Education Research: http://nieer.org/state-preschool-yearbooks/2018-2

Garcia, J.L., Heckman, J., Ermini Leaf, D., & Prados, M.J. (2017). Quantifying the life-cycle benefits of a prototypical early childhood program. Retrieved from https://heckmanequation.org/www/assets/2017/12/abc_comprehensivecba_JPE-SUBMISSION_2017-05-26a_sjs_sjs.pdf

General Accounting Office. (1998). Head Start: Challenges in monitoring program quality and demonstrating results. Washington, DC. U.S. General Accounting Office.

Gomby, D.S., Larner, M.B., Stevenson, C.S., Lewit, E.M., & Behrman, R.E. (1995). Long-term outcomes of early childhood programs: Analysis and recommendations. *The Future of Children*, 5(3), 6–24. doi:10.2307/1602365

Gomby, D.S., Krantzler, N., Larner, M.B., Stevenson, C.S., Terman, D.L., & Behrman, R.E. (1996). Financing child care: Analysis and recommendations. *The Future of Children*, 6(2), 5–25. doi:10.2307/1602416

Gormley Jr, W.T., Gayer, T., Phillips, D., & Dawson, B. (2005). The effects of universal pre-K on cognitive development. *Developmental Psychology*, 41(6), 872–884. doi:10.1037/0012-1649. 41.6doi:872

Gormley Jr, W.T., Phillips, D.A., Newmark, K., Welti, K., & Adelstein, S. (2011). Social-emotional effects of early childhood education programs in Tulsa. *Child Development*, 82(6), 2095–2109. doi:10.1111/j.1467-8624.2011.01648.x

Government Accountability Office. (2005). Head Start: Further development could allow results of new test to be used for decision making. Washington, DC:GAO.

Gray, S.W., & Klaus, R.A. (1965). An experimental preschool program for culturally deprived children. *Child Development*, 36, 887–898.

Heckman, J., Moon, S.H., Pinto, R., Savelyev, P., & Yavitz, A. (2010). The rate of return to the HighScope Perry Preschool Program. *Journal of Public Economics*, 94, 114–128.

Helburn, S., Culkin, M., Morris, J., Mocan, N., Howes, C., Phillipsen, L., … Rustici, J. (1995). Cost, quality, and child outcomes in child care centers: Public Report. Retrieved from https://files.eric.ed.gov/fulltext/ED386297.pdf#page=383

Hunt, J.M. (1961). *Intelligence and experience*. New York: Ronald Press Company.

Immelmann, K., & Soumi, S.J. (1982). Sensitive phases in development. In K. Immelmann, G.W. Barlow, L. Petrinovich, & M.B. Main (Eds.), *Behavioral development: The Bielefeld Interdisciplinary Project*. New York: Cambridge University Press.

Institute of MedicineandNational Research Council. (2015). Transforming the workforce for children birth through 8: A unifying foundation. Washington, DC: The National Academies Press.

Jensen, A. (1969). How much can we boost IQ and scholastic achievement? *Harvard Educational Review*, 39(1).

Jones, J. (2018). US early childhood policy: Towards a more coherent early childhood policy in the US. In L. Miller, C. Cameron, C. Dalli, & N. Barbour (Eds.), *The Sage handbook of early childhood policy* (pp. 133–150). London: Sage.

Kagan, S.L., Gomez, R., & Roth, J.L. (2018). Creating a new era of usable knowledge: Enhancing early childhood development through systems research. In L. Miller, C. Cameron, C. Dalli, & N. Barbour (Eds.), *The Sage handbook of early childhood policy* (pp. 566–580). London: Sage.

Karoly, L., Greenwood, P., Everingham, S., Hoube, J., Kilburn, M.R., Rydell, C.P., … Chiesa, J. (1998). Investing in our children: What we know and don't know about the costs and benefits of early childhood interventions. Retrieved from RAND: https://www.rand.org/pubs/monograph_reports/MR898.html

Kontos, S., Hsu, H.C., & Dunn, L. (1994). Children's cognitive and social competence in child care centers and family day-care homes. *Journal of Applied Developmental Psychology*, 15(3), 387–411. doi:10.1016/0193-3973(94)90039-6

Kontos, S.J. (1991). Child care quality, family background, and children's development. *Early Childhood Research Quarterly*, 6(2), 249–262. doi:10.1016/0885-2006(91)90011-9

Lazar, I., & Darlington, R.B. (1982). Lasting effects of early education: A report from the Consortium for Longitudinal Studies. *Monographs of the Society for Research in Child Development*, 47(2–3), 1–151. doi:10.2307/1165938

Lazarin, M., & McAdam, M. (2003). Comments on the Implementation of the National Head Start Reporting System on Child Outcomes. Public Comment. Hispanic Education Coalition, National Council of La Raza, National HEP-CAMP Association.

Lee, V.E., Brooks-Gunn, J., & Schnur, E. (1988). Does Head Start work? A 1-year follow-up comparison of disadvantaged children attending Head Start, no preschool, and other preschool programs. *Developmental Psychology*, 24(2), 210–222. doi:10.1037/0012-1649. 24. 2.doi:210

Lee, V.E., Brooks-Gunn, J., Schnur, E., & Liaw, F.-r. (1990). Are Head Start effects sustained? A longitudinal follow-up comparison of disadvantaged children attending Head Start, no preschool, and other preschool programs. *Child Development*, 61(2), 495–507. doi:10.2307/1131110

Long, S.K., & Clark, S.J. (1997). The New Child Care Block Grant: State funding choices and their implications. Retrieved from Urban Institute: https://www.urban.org/research/publication/new-child-care-block-grant

Matthews, H., Schulman, K., Vogtman, J., Johnson-Staub, C., & Blank, H. (2015). Implementing the Child Care and Development Block grant reauthorization: A guide for states. Washington DC: National Women's Law Centre and CLASP.

McCartney, K., Scarr, S., Phillips, D., & Grajek, S. (1985). Day care as intervention: Comparisons of varying quality programs. *Journal of Applied Developmental Psychology*, 6(2–3), 247–260. doi:10.1016/0193-3973(85)90061-9

McKey, R., Condelli, L., Ganson, H., Barrett, B., McConkey, C., & Plantz, M. (1985). The impact of Head Start on children, families, and communities: Final report of the Head Start evaluation, synthesis, and utilization project. Washington DC: U.S. Government Printing Office.

Meisels, S.J., & Atkins-Burnett, S. (2004). The Head Start National Reporting System: A Critique. *Beyond the Journal: Young Children*. Retrieved from http://journal.naeyc.org/btj/200401/meisels.pdf

Mitchell, A. (2001). *Prekindergarten programs in the States: Trends and issues*. Retrieved from Alliance for Early Childhood Finance: http://www.earlychildhoodfinance.org/downloads/2001/PreKStatePolicyTrends_final_2001.pdf

Moore, E., & Yzaguirre, R. (2004). Head Start's national reporting system fails our children: Here's why. *Education Week*, 23, 40–41.

National Academies of Sciences, Engineering, Medicine. (2018). Transforming the financing of early care and education. Washington, DC: National Academies Press. doi:10.17226/24984

National Research Council. (2008). Early childhood assessment: Why, what, and how. Washington, DC: The National Academies Press.

National Research Council and Institute of Medicine. (2000). From neurons to neighborhoods; the science of early childhood development. Washington, DC: National Academy Press.

Obama, B. (2013). The 2013 Presidential State of the Union address. Available at: https://obamawhitehouse.archives.gov/the-press-office/2013/02/12/remarks-president-state-union-address

Office of the Assistant Secretary for Planning and Evaluation. (1996). The personal responsibility and work responsibility reconciliation act of 1996. Retrieved from https://aspe.hhs.gov/report/personal-responsibility-and-work-opportunity-reconciliation-act-1996

Paulsell, D., Gordon, A., Nogales, R., Del Grosso, P., Sprachman, S., & Tarullo, L. (2006). Implementation of the Head Start National Reporting System: Spring 2005 update. Retrieved from ERIC: https://files.eric.ed.gov/fulltext/ED498080.pdf

Peisner-Feinberg, E.S., Burchinal, M.R., Clifford, R.M., Culkin, M.L., Howes, C., Kagan, S.L., … Zelazo, J. (2000). The children of the cost, quality, and outcomes study go to school: Technical report. Retrieved from Frank Porter Graham Child Development Center: https://fpg.unc.edu/sites/fpg.unc.edu/files/resources/reports-and-policy-briefs/NCEDL_CQO_executive_summary.pdf

Phillips, D., Gormley, W.T., & Anderson, S. (2016). The effects of Tulsa's cap head start program on middle-school academic outcomes and progress. *Developmental Psychology*, 52(8), 1247–1261. doi:10.1037/dev0000151

Phillips, D., Lipsey, M., Dodge, K., Haskins, R., Bassok, D., Burchinal, M., … Weiland, C. (2017). Puzzling it out: The current state of scientific knowledge on pre-K effects: A consensus statement. In *The current state of scientific knowledge on pre-kindergarten effects* (pp. 19–30). Washington, DC: Brookings Institution and Duke Center for Child and Family Policy.

Phillips, D., McCartney, K., & Scarr, S. (1987). Child-care quality and children's social development. *Developmental Psychology*, 23(4), 537–543. doi:10.1037/0012-1649. 23. 4doi:537

Ramey, C.T., & Campbell, F.A. (1984). Preventive education for high-risk children: cognitive consequences of the Carolina Abecedarian Project. *Am J Ment Defic*, 88(5), 515–523.

Ramey, C.T., Campbell, F.A., Burchinal, M., Skinner, M.L., Gardner, D.M., & Ramey, S.L. (2000). Persistent effects of early intervention on high-risk children and their mothers. *Applied Developmental Science*, 4, 2–14.

Ramey, C.T., & Ramey, S.L. (1998). Early intervention and early experience. *American Psychologist*, 53 (2), 109–120. doi:10.1037/0003-066X.53. 2. 109

Rebell, M.A., Wolff, J.R., Kolben, N., & Holcomb, B. (2017). Establishing universal access to prekindergarten as a constitutional right. Retrieved from Center for Educational Equity, Teachers College, Columbia University: http://www.centerforeducationalequity.org/publications/universal-prekindergarten-in-ny/Prekindergarten-as-a-Right_LongPaper_4B.pdf

Reynolds, A.J. (2000). *Success in early intervention: The Chicago child–parent centers*. Lincoln, NE: University of Nebraska Press.

Reynolds, A.J., Mann, E., Miedel, W., & Smokowski, P. (1997). The state of early childhood intervention: Effectiveness, myths and realities, new directions. *Focus*, 19(1), 5–11.

Reynolds, A.J., Ou, S.-R., & Temple, J.A. (2018). A multicomponent, preschool to third grade preventive intervention and educational attainment at 35 years of age. *JAMA Pediatrics*, 172(3), 247–256.

Reynolds, A.J., & Wolfe, B. (1997). Early intervention, school achievement, and special education placement: Findings from the Chicago Longitudinal Study. *Focus*, 19(3), 25–28.

Roth, W. (1976). The politics of daycare: The comprehensive Child Development Act of 1971. Retrieved from Institute for Research on Poverty, University of Wisconsin at Madison: https://www.irp.wisc.edu/publications/dps/pdfs/dp36976.pdf

Ruhm, C.J. (2011). Policies to assist parents with young children. *The Future of Children*, 21(2), 37–68.

Schulman, K., Blank, H., & Ewen, D. (1999). Seeds of success: State prekindergarten initiatives 1998–199. Washington, DC: Children's Defense Fund.

Senate-1512. (1971). Comprehensive child development act of 1971. Washington, DC: U.S. Government Printing Office.

Shepard, L., Kagan, S.L., & Wurtz, E. (1998). *Principles and recommendations for early childhood assessments*. Washington, DC: National Education Goals Panel.

Silver, S., & Sanders, S. (1997). *Learning, achieving, and developing by directing educational resources*. New York: New York State Assembly.

Smith, M., & Bissell, J.S. (1970). Report analysis: the impact of Head Start. *Harvard Educational Review*, 40(1), 50–105.

Sneha, E., Garcia, J.L., Heckman, J., & Hojman, A. (2015). Early childhood education. Retrieved from https://heckmanequation.org/contact/

Stevens, K.B. (2015). Renewing Childhood's Promise: The history and future of Federal Early Care and Education Policy. Washington, DC: American Enterprise Institute for Public Policy Research.

U.S. Bureau of Labor Statistics. (2009). Retrieved from U.S. Bureau of Labor Statistics: https://www.bls.gov/opub/mlr/2009/

Vogel, C., Nogales, R., Aikens, N., & Tarullo, L. (2008). Implementation of the Head Start National Reporting System: Spring 2006 (6094–6710). Retrieved from Mathematica: https://www.mathematica-mpr.com/our-publications-and-findings/publications/implementation-of-the-head-start-national-reporting-system-spring-2006

Weiland, C., & Yoshikawa, H. (2013). Impacts of a prekindergarten program on children's mathematics, language, literacy, executive function, and emotional skills. *Child Developement*, 84(6), 2112–2130. doi:10.1111/cdev.12099

Wu, P., & Campbell, D.T. (1996). Extending latent variable LISREL analyses of the 1969 Westinghouse head start evaluation to Blacks and full year Whites. *Evaluation and Program Planning*, 19(3), 183–191.

Yoshikawa, H. (1995). Long-term effects of early childhood programs on social outcomes and delinquency. *The Future of Children*, 5(3), 51–75. doi:10.2307/1602367

Yoshikawa, H., Weiland, C., Brooks-Gunn, J., Burchinal, M., Espinosa, L., Gormley, W.T., ... Zaslow, M. (2013). *Investing in our future: The evidence base on preschool education*. Retrieved from http://fcd-us.org/resources/evidence-base-preschool

Zigler, E. (2011). A warning against exaggerating the benefits of preschool education programs. In E. Zigler, W.S. Gilliam, & W. S. Barnett (Eds.), *The pre-K debates: Current controversies and issues* (pp. 197–200). Baltimore: Paul H. Brookes Publishing Co.

Zigler, E., & Berman, W. (1983). Discerning the future of early childhood intervention. *American Psychologist*, 38(8), 894–906. doi:10.1037/0003-066X.38. 8. 894

Zigler, E., Styfco, S.J., & Gilman, E. (1993). The national Head Start program for disadvantaged preschoolers. In *Head Start and beyond: A national plan for extended childhood intervention*. (pp. 1–41). New Haven, CT: Yale University Press.

Zigler, E., & Valentine, J. (Eds.). (1979). *Project Head Start: A legacy of the war on poverty*. New York, NY: The Free Press.

4

BIOECOLOGICAL SYSTEMS INFLUENCES ON EARLY CHILDHOOD EDUCATION

Jessica Navarro, Fabienne Doucet and Jonathan Tudge

Introduction

Urie Bronfenbrenner's "theoretical paradigm ... has transformed the way many social and behavioral scientists approach, think about, and study human beings and their environments" (Moen, Elder, & Luscher, 1995, p. 1). Since its introduction in 1979, Bronfenbrenner's theory has, indeed, stimulated a considerable amount of discussion and debate. Its evolution over the years from ecological theory to bioecological paradigm (see Bronfenbrenner, 1979b, 1989a, 1995a; Bronfenbrenner & Ceci, 1994; Bronfenbrenner & Morris, 1998; Rosa & Tudge, 2013), and the introduction of the Process-Person-Context-Time (PPCT) model (Bronfenbrenner, 1995b, 2001; Bronfenbrenner & Morris, 1998) have spurred interest in reconceptualizations of person–environment interrelatedness.

The importance of environment to the understanding of human development had long been argued in the classic nature–nurture debate, but Bronfenbrenner's approach to environment had a different tone because it extended beyond showing how variations in environment could alter genetic development. Bronfenbrenner's model emphasized a more complex understanding of environment as more than the immediate, tangible surroundings of a developing individual (Bronfenbrenner, 1979b). Instead, Bronfenbrenner explained environment, or context, to be composed of multiple layers. For many scholars and practitioners, this more sophisticated understanding of context is where Bronfenbrenner's theory ends, though in fact it is only where it begins.

In his later work, Bronfenbrenner articulated that proximal processes (the everyday reciprocal interactions between a developing child and his or her environments) are the "engines of development" (Bronfenbrenner & Evans, 2000, p. 118). Similarly, it could be argued that Urie Bronfenbrenner was an "engine" in the development of early childhood education (ECE) in the United States. Throughout his long and distinguished career Bronfenbrenner contributed to modern ECE in a multitude of ways. Bronfenbrenner's theoretical model of development provided a framework for not only academic research, but also for ECE practitioners and administrators. Further, Bronfenbrenner was instrumental in the development of the national Head Start program (Hayes, O'Toole, & Halpenny, 2017). Before exploring how Urie Bronfenbrenner's theories of human development influence and contribute to the field of ECE, we will explore the development of Bronfenbrenner's theory across his career, ending with an in-depth discussion of the bioecological model of human development, and the importance of proximal processes and the PPCT model for early childhood education.

Biographical Background

Urie was born in Russia in 1917 to Alexander and Eugenie Bronfenbrenner. Alexander, a neuropathologist, and his wife Eugenie were both Jewish, and while they were part of the professional class they did not feel secure in revolutionary Russia. Alexander emigrated to the United States in 1921, and Eugenie and Urie followed in 1923. Alexander secured a position leading a laboratory in upstate New York, where he worked with people with mental disabilities. Young Urie often helped his father in the lab and developed an interest in psychology (Bronfenbrenner, 1995; Hayes et al., 2017; Tudge, Merçon-Vargas, Liang, & Payir, 2017). Inspired by discussions with his father and using his own observations, Urie began to think about bidirectional relationships between the organisms and their environment, which would later become a core idea of his bioecological model of human development.

Bronfenbrenner earned undergraduate degrees in psychology and music from Cornell, and then went on to earn his Master's from Harvard and Doctorate from the University of Michigan. During World War II Bronfenbrenner served as a psychologist at military hospitals and returned to join the faculty at Cornell in the Child Development and Family Relations department, where he remained for the duration of his career until his retirement and subsequent death in 2005.

Bronfenbrenner's contributions to the field of child development and early childhood education spanned myriad fields of interest and topics (Tudge, 2017). He wrote about cross-cultural studies on children and their families in Russia and the United States (Bronfenbrenner, 1971), and was particularly interested in how social changes related to urbanization and changing family structures impacted child development in these different societies. These writings informed the development of his ecological (later termed "bioecological") theory, as well as his involvement in public policy related to child welfare, family services, and state-supported early childhood education. During the Johnson administration, Bronfenbrenner served on a committee to develop a nationwide, federally funded program to assist young children from poor families before entering elementary school. This program, Head Start, was launched in 1965 (Hayes et al., 2017).

Throughout his extensive career, Bronfenbrenner's early ideas about the bidirectional and dynamic interplay between people and their environments continued to be central to the various iterations of his theory of human development. This is evident in the use of the word "ecology" throughout his career. Ecology is defined as the interrelations between people and the environments in which they live. Bronfenbrenner initially termed his theory "the ecology of human development" before modifying it to "ecological systems theory" and finally to "bioecological theory." These iterations were developed in phases over the course of his career, as outlined in our previous work (Rosa & Tudge, 2013). While these phases are distinct and show an evolution of his theory, they all share the same underlying tenet of ecology (Tudge, Payir & Merçon-Vargas, under review; Tudge, Rosa, & Payir, 2018).

Part I: The Evolution of the Bioecological Model of Human Development

A self-reflective theorist, Bronfenbrenner identified three phases in the development of his theory. Phase 1 began in 1973 and concluded with the publication of *The Ecology of Human*

Development in 1979. This phase focused largely on the differing contexts in which developing individuals are situated. Phase 2 (1983–1993) expanded on these ideas, placing greater emphasis on the role of the individual (including person characteristics) and the role of time in development. During Phase 2 Bronfenbrenner also presented initial models of human development, including the person-process-context model (1983–1986) and the process-person-context model (1988–1989). In the final phase, occurring between 1993 and 2006, Bronfenbrenner defined proximal processes and the Process-Person-Context-Time (PPCT) model for research. These phases can also be linked to some of his most influential publications (Bronfenbrenner & Evans, 2000): his 1979 book during the first phase, and two "reformulations" (Bronfenbrenner & Crouter, 1983 and Bronfenbrenner & Morris, 1998) demarcating phases two and three. As noted in our previous work (Tudge et al., 2009; Tudge et al., 2016), many scholars rely upon earlier versions of Bronfenbrenner's theory that he himself later criticized, but, as we will argue, this shortsighted approach fails to do justice to the developed form of the theory.

While we will explore the development of his theory across these three phases, we encourage researchers and practitioners alike to focus on and utilize the final form of his theorizing (the bioecological theory and PPCT model described in Phase 3).

Phase 1 The Ecology of Human Development (1973–1979)

Inspired by his research with children and families and his experiences in social policy, Bronfenbrenner began his theoretical work to address two issues: (1) the limitations of laboratory-based psychology research in real-world applications, and (2) the resulting dearth of information applicable to the formation of social policies and programs for children and their families. He wrote:

> Our science is peculiarly one-sided. We know much more about children than about the environments in which they live or the processes through which these environments affect the course of development … [This largely results from] the absence of a theoretical framework appropriate for analyzing the environments in which human beings live.
>
> *(Bronfenbrenner, 1979a, p. 844)*

Bronfenbrenner believed that lab-based research and the prevailing theories of child development did not consider the contexts in which children and their families actually lived and interacted, and were therefore not helpful in the formation of social policies to support families. Contrary to beliefs at the time, Bronfenbrenner argued that research should be guided by social policy, therefore attempting to provide useful answers to issues of the day (Rosa & Tudge, 2013).

In articles throughout the 1970s, and culminating in his 1979 monograph, Bronfenbrenner outlined the first iteration of his theory. He defined the ecology of human development as:

> the scientific study of the progressive, mutual accommodation between an active, growing human being and the changing properties of the immediate settings in which the developing person lives, as this process is affected by relations between these settings, and by the larger contexts within which the settings are embedded.
>
> *(Bronfenbrenner, 1979b, p. 21)*

Put simply, Bronfenbrenner viewed development as being influenced by the interrelations and interactions between individuals and the different levels of contexts in which they lived. In this monograph, Bronfenbrenner described the different "systems" of context as "a set of nested structures, each inside the other like a set of Russian dolls" (Bronfenbrenner, 1979b, p. 3).

With the developing individual (e.g., a child) at the center, the ecological system was composed of the microsystem, the mesosystem, the exosystem, and the macrosystem. We discuss each of these systems more fully later in the chapter but note here that Bronfenbrenner's concept of nested systems became immensely popular after the release of his 1979 monograph. Textbooks included diagrams created to visually represent these levels, which were depicted as concentric circles. These diagrams and the "Russian doll" metaphor offered an appealing simplicity to explain the numerous external factors impacting child development. However, many scholars failed to recognize the bidirectionality of his theory or apply it to more than one level of context. Researchers cited Bronfenbrenner's ecological system theory when studying children in a real-world microsystems (e.g., school or home), but often overlooked four key points: (a) children interact differently based on the context, influencing what appear to be individual characteristics, (b) children and microsystems interact and influence one another in a bidirectional relationship, (c) macrosystems continually influence the microsystems children inhabit and (d) microsystems are not objective realities but need to be understood as they are perceived by those within them. Bronfenbrenner clearly articulated this final point with his definition of ecological validity as:

> The extent to which the environment experienced by the subjects in a scientific investigation has the properties it is supposed or assumed to have by the investigator. … Again, the use of the term *experienced* in the definition highlights the importance of the phenomenological field in ecological research.
>
> *(Bronfenbrenner, 1979b, p. 29, italics in the original)*

Of particular interest to ECE researchers and practitioners, Bronfenbrenner introduced the term *ecological transition* during this phase of his career. Bronfenbrenner considered an ecological transition to be a change in or between systems prompted by individual developmental processes (e.g., a child matures to preschool readiness), changes in the environment (e.g., a place opens in an appropriate classroom for a child), or a combination of individual and environmental factors. Bronfenbrenner also conceived that transitions could be the motive for development, such as the development of more independence as a result of the transition to school (Doucet, 2008). Furthermore, for Bronfenbrenner, ecological transitions impacted not only the individual, but the systems they inhabit and with which they interact (e.g., family, friends) (Rosa & Tudge, 2013). The richness of investigating the impact of ecological transitions on developmental trajectories was expressed by Bronfenbrenner and Crouter (1983): "From the point of view of scientific method, every ecological transition has the virtue that it constitutes a readymade experiment of nature with a built-in, before/after design in which each subject can serve as his [sic] own control" (p. 381). As we will discuss in the second part of this chapter, an understanding of ecological transitions can help inform and support both children and their families during times of educational transition (e.g., from preschool to kindergarten) (Doucet, 2008; Doucet & Tudge, 2007; Hayes et al., 2017; Tudge, Freitas, & Doucet, 2009).

Thus, although Bronfenbrenner's work from the 1970s is often reduced to a theory of how contexts impact individuals, the underlying tenet of these ideas was ecological – human development occurs through complex, dynamic, and interdependent interactions between individuals and their environment (Tudge et al., 2018).

Phase 2 Ecological Systems Theory (1983–1993)

Bronfenbrenner himself observed the simplification of his 1979 monograph to a treatise on context and wrote in the 1980s and early 1990s to emphasize the ecological nature of his theory. During this phase, Bronfenbrenner also expanded on his theory; he further delineated the role person characteristics and time play in development and elaborated on the role culture plays in the macrosystem. Bronfenbrenner also introduced his first models for research during this phase, the person-process-context model (1983–1986) and the process-person-context model (1988–1989).

Bronfenbrenner expounded upon the role of individuals in their own development, namely two types of "instigative characteristics." These two types include "personal-stimulus qualities," expressed as the immediate reactions of the individual to either welcome or discourage interactions with their environment, and "developmentally structuring personal attributes," which Bronfenbrenner defined as "modes of behavior or belief that reflect an active, selective, structuring orientation toward the environment and/or tend to provoke reactions from the environment" (Bronfenbrenner, 1989a, p. 223). Personal characteristics were further refined during the third phase.

During this period, Bronfenbrenner also began to use the term "chronosystem" to describe the influence of time on human development. Throughout the life course, a multitude of events and experiences could change both internal and external factors that could promote or prohibit development. Further, Bronfenbrenner argued that the chronosystem could be used to describe changes not only in the individual, but also changes in contextual systems. Bronfenbrenner's subsequent elaborations of the concept of chronosystem are discussed in more detail later in the chapter.

Bronfenbrenner's writings also emphasized variations across cultures and sub-cultures, and how these variations impact human development. Inspired by Vygotsky, Bronfenbrenner wrote that "from earliest childhood onward the development of one's characteristics as a person depends in significant degree on the options that are available in a given culture at a given point in its history" (Bronfenbrenner, 1989a, p. 228). During this period, Bronfenbrenner also highlighted the importance of belief systems, as they form the scaffold from which parents, teachers, and other caregivers interact with and socialize children.

Phase 3 Bioecological Theory (1993–2006)

During this phase, Bronfenbrenner added the prefix "bio" to the name of the theory, perhaps in part to emphasize the often overlooked importance of person ("bio") characteristics in development. This period, which stretched from the early 1990s to the publishing of a posthumous chapter in 2006, remained true to his earlier emphasis on ecology, but added a fundamental idea – proximal processes. Proximal processes are the everyday activities and interactions between the developing individuals and their environments. Bronfenbrenner believed that proximal processes are the "engines of development" (Bronfenbrenner & Evans, 2000, p. 118) and placed them at the forefront of the Process-Person-Context-Time (PPCT) model. In this final iteration of the model, Bronfenbrenner continued to view individual characteristics and contexts as key elements of development, but secondary to proximal processes. During this phase Bronfenbrenner also further elaborated on time as a driving force in development, as well as time as a necessary component of research designs. To study human development over time, Bronfenbrenner argued that one must collect data longitudinally.

As is evident from our overview of the three phases of Bronfenbrenner's "ecological" theory, his ideas underwent major revisions throughout his career, culminating in the bioecological theory and PPCT model. As this final iteration of his theory encompasses processes, person, context, and time, we recommend that researchers and practitioners alike embrace this final model as opposed to earlier versions. As Rosa and Tudge (2013, p. 256) stated:

> [T]here really is no reason for continuing to treat Bronfenbrenner's theory as one of contextual influences on development, or for ignoring the focus, during the third and final phase, on proximal processes and the use of the PPCT model as a guide for research using the bioecological theory.

Part II:The PPCT Model

Rather than being discrete variables, all four components of the PPCT model function interdependently and synergistically. While it may seem overwhelming to contemplate how this complex and dynamic model shapes child development (and thus research or practice), there is also great power in it – this model provides a scaffold for beginning to understand the interconnected forces at work in human development. This approach acknowledges that children do not develop in isolation, but in real families, in real schools, and in the real world. As proximal processes form the heart of this model, this is where we will begin our discussion.

Proximal Processes

As noted previously, proximal processes are the increasingly complex reciprocal interactions taking place between the child and their environment, including the people, objects, and symbols they engage with. Bronfenbrenner and Morris (1998, p.996) elucidated:

> Especially in its early phases, but also throughout the life course, human development takes place through processes of progressively more complex reciprocal interaction between an active, evolving biopsychological human organism and the persons, objects, and symbols in its immediate environment. To be effective, the interaction must occur on a fairly regular basis over extended periods of time.

In a speech to UNESCO, Bronfenbrenner used an apt analogy to describe increasingly complex reciprocal interactions:

It's what happens in a ping-pong game between two players as the game gets going. As the partners become familiar with each other, they adapt to each other's style. The game starts to go faster, and the shots in both directions tend to become more complicated, as each player, in effect, challenges the other.

(Bronfenbrenner, 1989b)

Bronfenbrenner and Morris (2006, p. 996) provided examples of these types of interactions:

> Feeding or comforting a baby, playing with a young child, child–child activities, group or solitary play, reading, learning new skills, athletic activities, problem-solving, caring

for others in distress, making plans, performing complex tasks, and acquiring new knowledge and know-how.

These examples encompass myriad activities taking place daily in the home of a young child or at an early childhood education center. Bronfenbrenner also viewed proximal processes as being positive interactions, that is, interactions that would promote capacity building regardless of situation. In other words, proximal processes could both promote development in a child already doing well or act as a protective force for a child facing challenges.

As articulated above, proximal processes are not a stand-alone component of the PPCT model. They are synergistically influenced by the individual characteristics of the developing individual, the multitude of contexts in which that individual interacts, and time:

> The form, power, content, and direction of the proximal processes effecting development vary systematically as a joint function of the characteristics of the *developing person*; of the *environment* – both immediate and more remote – in which the processes are taking place; the nature of the *developmental outcomes* under consideration; and the social continuities and changes occurring over *time* through the life course and the historical period during which the person has lived.
>
> *(Bronfenbrenner & Morris, 1998, p. 996, italics in the original)*

These synergistic forces push development forward by continually shaping and changing one another. For example, the proximal processes experienced by an infant (e.g., a strong parent–infant relationship) impact the development of individual characteristics (e.g., internalization of a secure attachment), which impacts later proximal processes (e.g., little separation anxiety at drop-off at child care). These forces do not operate in "nested" contexts as Bronfenbrenner initially described, but in a more complex and dynamic network (Tudge, 2008).

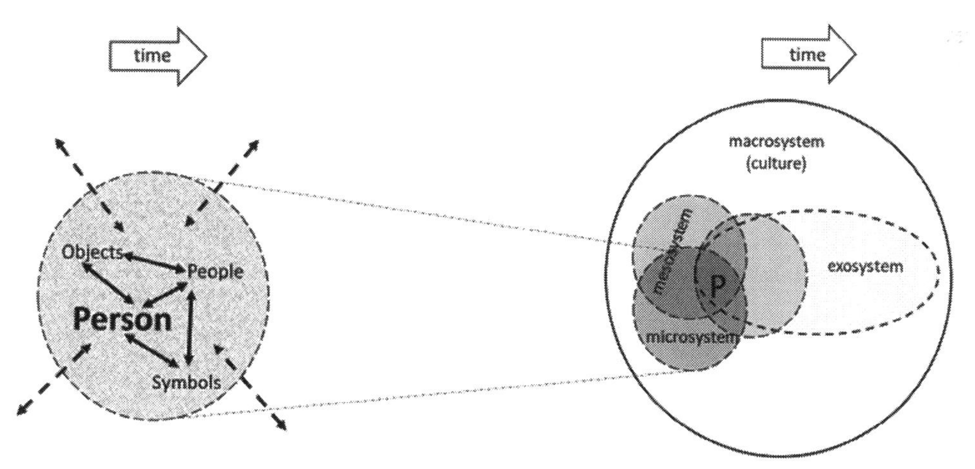

The developing person, engaging in proximal processes with symbols, objects and other people, within a permeable microsystem.

Proximal processes both influence and are influenced simultaneously by characteristics of the people involved, the contexts in which they are situated, and time.

The microsystem is linked with others in the mesosystem, with the exosystem and macrosystem indirectly influencing (and being influenced by) the developing person.

FIGURE 4.1 Visual Representation of the PPCT Model of Urie Bronfenbrenner's Bioecological Theory
Source: Adapted from Tudge (2008).

Discussions of "quality" in early childhood education often focus on the built environment, resources, curriculum, and adequate staffing. When using the lens of proximal processes as the driving force of development, it is clear that the everyday interactions and activities in which a child engages are of critical importance to enhancing quality and realizing the goals of early childhood education. Positive teacher–child, teacher–parent, and child–peer relationships are the key to promoting proximal processes, and it is vital that these relationships are supported and cultivated in an early childhood setting (Hayes et al., 2017). Strong emotional relationships at school have the power to help young children achieve "competence" as well as overcome challenges and obstacles in other areas of their lives (Bronfenbrenner, 2001; Rosa & Tudge, 2013).

So how should early childhood educators go about providing these progressively more complex interactions to help their students develop toward competence? One clue to this question can be derived from the definition of competence as the "demonstrated acquisition and further development of knowledge, skills or ability to conduct and direct one's own behavior across situations and developmental domains" (Bronfenbrenner & Evans, 2000, p. 118). The other clue is from the model itself; where the developing individual of interest (e.g., child or teacher) takes center stage. We contend that early childhood educators should adopt a child-oriented paradigm and help *guide* the child to be able to "direct one's own" interactions and activities. In an early paper that delineated principles for early childhood care, Bronfenbrenner (1971) stated that "the program should be designed so as to allow ample opportunity for the child to observe, ask questions, experiment, and search out answers to increasingly complex problems" (p. 89). Further, Bronfenbrenner (1989a) emphasized that teaching between teacher and student is a bidirectional process: "By now it should be clear that what is made possible by 'progressively more complex reciprocal interaction' is a process of mutual 'education;' the child is 'teaching' the adult, and the adult is 'teaching' the child."

Peer interactions also play an important role in the development of competence in myriad ways, including the development of social and linguistic skills. Peer relationships provide an opportunity for play where children can "explore the boundaries of their world – boundaries typically drawn by adults" (Hayes et al., 2017, p. 89). Pretend play allows children to "write the script for endless situations, challenge the logic of adult rules and develop their own internal logic" (Hayes et al., 2017, p. 90). These play scenarios become increasingly complex through time, as children themselves mature and face new challenges.

In addition to reciprocal interactions with other persons, proximal processes take place between developing individuals and objects and symbols in their environments. Bronfenbrenner and Morris (2006) wrote that "the objects and symbols in the immediate environment must be of a kind that invites attention, exploration, manipulation, elaboration, and imagination" (p. 798). Objects and symbols in the classroom (or in an outdoor play area) that stimulate repeated and progressively complex interactions are likely to have a greater impact on development than those that do not invite such interactions.

Person Characteristics

In the third phase of his career Bronfenbrenner elaborated on previous notions of the role individual characteristics play in development. Formerly defined as "personal-stimulus qualities" and "developmentally structuring personal attributes," Bronfenbrenner and Morris (1998) refined person characteristics into three categories: *demand, resource*, and *force*. These characteristics are not static, but dynamic qualities that constantly evolve and change as a result of interrelated systems and interactions (i.e., proximal processes, context and time).

Individual characteristics are part of the spiral of development, where they are both the result of proximal processes and an influence on these interactions and systems:

> Qualities of the developing person ...emerge at a later point in time as the result of the joint, interactive, mutually reinforcing effects of the four principal antecedent components of the model. In sum, in the bioecological model, the characteristics of the person function both as an indirect producer and as a product of development.
>
> *(Bronfenbrenner & Morris, 1998, p. 996)*

Demand

Demand characteristics are qualities that are immediately apparent and serve to either encourage or discourage social interactions. These initial impressions may include physical attributes like gender, skin color, and age, as well as observable attitudes and behaviors like shyness, anxiety, or happiness. In an early childhood setting these initial impressions are formed by all the individuals in the system – teachers, parents, and children.

The demand characteristics of the early childhood practitioner play an initial role in how children interact with their environments. For example, a young child may observe that her teacher looks kind or worried, or that her teacher has a different skin color from her, or that her teacher appears younger than her parents. All of these observations influence the initial interactions between a child and his or her teacher. Similarly, the first impression teachers make of new students can greatly influence how they initially interact and the type of proximal processes that can result. These demand characteristics can be helpful (e.g., a young boy clinging to his father can signal separation anxiety and indicate that the teacher should approach the child warmly and with reassurance), but they can also be misleading (e.g., a young child first appears to be hyperactive and agitated, but this subsides as she becomes more comfortable in her new surroundings). These initial misinterpretations can negatively influence interactions, leading to fewer opportunities for proximal processes, and block the formation of positive teacher–child relationships. Early childhood practitioners should be self-reflexive and question their own assumptions (e.g., gender and class stereotyping, developmental guidelines) until they have observed and interacted with the child extensively – when they have a better understanding of the resource and force characteristics of that individual child.

Resource

Resource characteristics are not immediately apparent; they are "biopsychological liabilities and assets that influence the capacity of the organism to engage effectively in proximal processes" (Bronfenbrenner & Morris, 2006, p. 812). These "assets" can include abilities, skills, prior experiences, knowledge, or a secure parental attachment, all of which promote proximal processes and development toward competence. "Liabilities" limit the extent to which the developing child can engage in proximal processes. These may include illness, physical or mental disabilities, a chaotic home environment, or other social impairments. Through observations and interactions, impressions based on demand characteristics may be altered to reflect resource characteristics.

Early childhood administrators and practitioners can assist children to engage in proximal processes by helping to ameliorate disruptive resource characteristics. For example, a young boy with an articulation disorder may have more positive proximal processes if his teacher is given the support and knowledge (e.g., a family meeting with parents and a speech therapist to make sure all parties are using similar terminology and techniques) to accommodate this limitation.

Force

Force characteristics "involve such active orientations as curiosity, tendency to initiate and engage in activity alone or with others, responsiveness to initiatives by others, and readiness to defer immediate gratification to pursue long-term goals" (Bronfenbrenner & Morris, 1998, p. 1009). Like resource characteristics, force characteristics can also be disruptive, and include "impulsiveness, explosiveness, distractibility, [and] inability to defer gratification" (p. 1009).

As mentioned previously, these individual characteristics do not operate in isolation. Other people (including teachers, parents, family, and peers) who interact with the developing child have their own demand, resource, and force characteristics. These characteristics also influence the interactions, relationships, and proximal processes of the child.

Context

In addition to person characteristics, proximal processes are also affected by the context of the developing individual. The importance of context was a constant throughout the evolution of Bronfenbrenner's theory of human development. Bronfenbrenner and Morris (2006) reiterated the 1979 metaphor that "the ecological environment is conceived of as a set of nested structures, each inside the other like a set of Russian dolls" (p. 814). However, the bioecological theory and PPCT model assert that context is only one of the factors that impact proximal processes, the "engines of development" (Bronfenbrenner & Evans, 2000, p. 118). All proximal processes take place in the microsystem, but are also impacted by "forces emanating from multiple settings and from the relations among these settings" (Bronfenbrenner & Morris, 2006, p. 816). We will now examine these settings and their importance to early childhood education, beginning with the most proximal – the microsystem.

Microsystem

> Bronfenbrenner (1994) defined the microsystem as "a pattern of activities, social roles, and interpersonal relations experienced by the developing person in a given face-to-face setting with particular physical, social, and symbolic features … in the immediate environment" (p. 1645). Young children engage in proximal processes in many settings (e.g., home, school, park, grandparents' home), but we will focus on two key microsystems relevant to this chapter: the home and early education centers.

Bronfenbrenner recommended that the physical environment should engage the developing child in activities than can be repeated with increasing difficulty, while at the same time minimizing distractions that undermine development (Bronfenbrenner, 1971). Further, Bronfenbrenner and Morris (2006) hypothesized: "Not only do developmentally generative features of the surroundings have greater impact in stable settings, but they also buffer against the disruptive influences of disorganizing environments" (p. 815). Several early childhood pedagogies place great importance on the physical environment, including the Reggio Emilia and Montessori approaches. Early childhood educators can envisage materials and activities to encourage proximal processes by perceiving the classroom from the eyes of their students

(Hayes et al., 2017). For example, what proximal processes might a child engage in while sitting at a desk and answering questions when called upon? What about a classroom with a multitude of developmentally appropriate materials from which the child can freely choose and discuss in small groups with their peers? Further, imagine a loud and hectic classroom without a predictable routine. What proximal processes might occur in this type of environment as opposed to "a context that is patterned, stable, and familiar" (Bronfenbrenner, 1971, p. 89)?

In addition to taking this child-centered approach to creating and organizing microsystems (including classrooms and outdoor play spaces), early childhood teachers should also incorporate their knowledge of child development to ensure proximal processes are encouraged for all students, regardless of ability. As all children develop at different paces, teachers should differentiate materials, activities, and interactions based on the skills and needs of the individual child, not the class as a whole (Bronfenbrenner, 1971).

Mesosystem

Bronfenbrenner and Morris (2006) defined a mesosystem as the "relationships existing between two or more settings; in short, it is a system of two or more microsystems" (p. 817). The mesosystem is an important component of context in the lives of young children because it allows for an analysis of the *transitions* between microsystems (e.g., drop-off and pick-up from school, moving from outdoor recess to indoor activities, the move from preschool to elementary school). The linkages (i.e., mesosystems) between home and school are of particular importance in early childhood education. Bronfenbrenner (1971) wrote extensively about the need for strong relations and communication between these two microsystems: "It is only by working jointly with the child *and his parents* that a day care program can achieve its objective of creating an environment that permits the realization of human potential" (p. 88, italics in the original).

To reduce stress related to the transition between home and school, Bronfenbrenner recommended that early care educators engage in positive proximal processes with both parent and child:

> It is the responsibility of the staff to assist the parents in this process by preparing them in advance, encouraging an appropriately paced withdrawal, and providing the necessary attention once the parents have left to enable the child to feel secure in the new setting.
>
> *(Bronfenbrenner, 1971, p. 89)*

Further, he recommended that schools proactively encourage parental participation and engagement with their children, both at school and home: "an effective day care program must create every possible opportunity for enhancing the amount and, especially, the quality of family interaction with children both in the day care setting and in the home" (Bronfenbrenner, 1971, p. 91).

In addition to supporting transitions between these two microsystems, early childhood administrators and educators must be cognizant of cultural and linguistic diversity within the families they serve. Bronfenbrenner (1971) wrote: "So that both parents and children can feel a sense of identity ... settings should not be cast in a single mold but reflect the particular characteristics of different regions and styles of life" (p. 93). Educators should be especially proactive in partnering with culturally and linguistically diverse families to ensure that relations between the home and the school are positive, thus promoting productive proximal processes in both settings (Hayes et al., 2017).

Mesosystem analysis can also be insightful in understanding why a child may be engaging in positive proximal processes in one setting but not another. One can imagine, for example, homes in which children may be encouraged to engage with their parents in relatively egalitarian ways but come to a school in which the teacher defines herself as the clear authority figure. By contrast, parents might insist on obedience to their rules at home but children might be in a school setting in which children are given some freedom to choose which of the available activities they want to engage in.

Exosystem

Exosystems have an *indirect* influence on the proximal processes of a developing child. They are activities and interactions taking place in microsystems in which the developing child is not engaged, but yet still influence the proximal processes of the child (Bronfenbrenner & Morris, 2006). For example, an educator's chaotic home life may reduce the quality of interactions he is able to engage in with his students, or the decisions about funding made at school board meetings may restrict the range and qualities of materials a student can engage with.

Bronfenbrenner (1971) recommended that administrators (operating in an exosystem of the developing child) should take special care in the hiring of *qualified* early childhood educators, as they are largely responsible for promoting productive proximal processes. He believed that such individuals could "create an atmosphere that is open and interesting, well-organized without being rigid – one in which people, adults and children, can enjoy and trust each other" (Bronfenbrenner, 1971, p. 92).

Macrosystem

Like exosystems, the influence of macrosystems is indirect, but they operate at a societal, cultural, or sub-cultural level. Bronfenbrenner (1989b) conceived of the macrosystem as a group of people who share the same culture or sub-culture – whose shared values, beliefs, practices, resources, and sense of identity unite them into a specific group (Bronfenbrenner, 1989a; Tudge, 2008). These cultural elements indirectly influence the microsystems of developing children, thus impacting their proximal processes. Bronfenbrenner (1989a) wrote: "It is from this repertoire [of culture-wide values and beliefs] that parents, teachers, and other agents of socialization draw when they, consciously or unconsciously, define the goals, risks, and ways of raising the next generation" (p. 228). For example, the child-rearing values of a particular culture may impact how parents socialize, discipline, and raise their children. In the United States, middle-class parents and working-class parents place different value on independence; middle-class parents allow their children to make more decisions on their own, while working-class parents tend to want their children to follow rules (Tudge et al., 2017). Early childhood educators should be alert to and cognizant of the varying cultures they serve.

In addition, the macrosystem also encompasses resources in the community (e.g., funding for early childhood programs, safety, access to healthy food and clean drinking water, transportation, etc.). An early education center in an affluent area may have the resources to build a space especially designed for young children, to hire qualified teachers, and to have smaller classes. In this instance, the resources of the macrosystem indirectly influence the proximal processes children engage in with other people, objects, and symbols.

Time

Time is the fourth component of the PPCT model and is subdivided into three different dimensions of time: microtime, mesotime, and macrotime (Bronfenbrenner & Morris, 2006). *Microtime* refers to the time passing while proximal processes are taking place. For example, the time during which a young child engages with pouring water back and forth between pitchers. As Hayes et al. (2017) aptly described: "The term microtime evokes this sense of attending to what is happening in the moment – the child's 'now' activities – rather than focusing on the outcome" (p. 107). For early childhood teachers, being cognizant of microtime allows a concentrated child to actively and purposefully engage with his or her environment without interruption, ultimately encouraging him or her to participate in proximal processes. *Mesotime* refers to the repetition of interactions and activities and is one of the defining properties of proximal processes: "the interaction must occur on a fairly regular basis over extended periods of time" (Bronfenbrenner & Morris, 1998). *Macrotime* refers to the historical time period in which the developing individual lives. Cultural beliefs, values, morals, and practices are dynamic across time and generations, even within the same population. For example, in the early part of the twentieth century most care for young children was provided by family members in the home. Post World War II, changing family systems created the need for more widespread early childhood education.

Conclusion

While theories of human development are often employed in developmental and educational research, they also have considerable benefits to offer practitioners in the field of ECE. Urie Bronfenbrenner's bioecological theory of human development and the Process-Person-Context-Time (PPCT) model offer a lens for understanding the complex and "linked" lives of children, families, and schools. Teachers play an important role in the lives of young children, partnering with parents to promote productive proximal processes and help children to succeed. Bronfenbrenner's theory is much more than nested circles of context – it ultimately places children and their interactions at the heart of development:

> In order to develop intellectually, emotionally, socially, and morally a child requires, for all of them, the same thing: participation in progressively more complex reciprocal activity, on a regular basis over an extended period in the child's life, with one or more other persons with whom the child develops a strong, mutual, irrational emotional attachment, and who is committed to the child's well-being and development, preferably for life.
>
> *(Bronfenbrenner, 1989a)*

References

Bronfenbrenner, U. (1971). Day care USA: A statement of principles. *Peabody Journal of Education*, 48(2), 86–95.

Bronfenbrenner, U. (1979a). Contexts of child rearing: Problems and prospects. *American Psychologist*, 34 (10), 844–850.

Bronfenbrenner, U. (1979b). *The ecology of human development: Experiments by nature and design*. Cambridge: Harvard University Press.

Bronfenbrenner, U. (1989a). Ecological systems theory. In R. Vasta (Ed.), *Annals of child development*, vol. 6 (pp. 187–249). Greenwich, CT: JAI Press.

Bronfenbrenner, U. (1989b). *Who cares for children?* (no page numbers). Paris: UNESCO.

Bronfenbrenner, U. (1995a). The bioecological model from a life course perspective: Reflections of a participant observer. In P. Moen, G.H. Elder, Jr., & K. Lüscher (Eds.), *Examining lives in context: Perspectives on the ecology of human development* (pp. 599–618). Washington, DC: American Psychological Association.

Bronfenbrenner, U. (1995b). Ecological models of human development. In M. Gauvin & M. Cole (Eds.), *Readings on the development of children* (pp. 3–8). New York: Worth Publishing.

Bronfenbrenner, U. (2001). The bioecological theory of human development. In N.J. Smelser & P.B. Baltes (Eds.), *International encyclopedia of the social and behavioral sciences*, vol. 10 (pp. 6963–6970). New York, NY: Elsevier.

Bronfenbrenner, U., & Ceci, S. (1994). Nature–nurture reconceptualized in developmental perspective: A bioecological model. *Psychological Review*, 101(4), 568–586.

Bronfenbrenner, U., & Crouter, A.C. (1983). The evolution of environmental models in developmental research. In P.H. Mussen (Series Ed.) & W. Kessen (Vol. Ed.), *Handbook of child psychology, Vol. 1: History, theory, methods* (4th ed., pp. 357–414). New York: Wiley.

Bronfenbrenner, U., & Evans, G.W. (2000). Developmental science in the 21st century: Emerging questions, theoretical models, research designs and empirical findings. *Social Development*, 9(1), 115–125.

Bronfenbrenner, U., & Morris, P.A. (1998). The ecology of developmental processes. In W. Damon (Series Ed.) & R.M. Lerner (Vol. Ed.), *Handbook of child psychology: Vol. 1. Theoretical models of human development* (5th ed., pp. 993–1028). New York: John Wiley.

Bronfenbrenner, U., & Morris, P.A. (2006). The bioecological model of human development. In W. Damon & R.M. Lerner (Series Eds.) & R.M. Lerner (Vol. Ed.), *Handbook of child psychology: Vol. 1. Theoretical models of human development* (6th ed., pp. 793–828). New York: John Wiley.

Doucet, F. (2008). How African American parents understand their and teachers' roles in children's schooling and what this means for preparing preservice teachers. *Journal of Early Childhood Teacher Education*, 29(2), 108–139. doi:10.1080/10901020802059441

Doucet, F., & Tudge, J.R.H. (2007). Co-constructing the transition to school: Reframing the "novice" versus "expert" roles of children, parents, and teachers from a cultural perspective. In R. Pianta, M. Cox, & K. Snow (Eds.), *School readiness and the transition to kindergarten in the era of accountability* (pp. 307–328). Baltimore, MD: Brookes Publishing.

Hayes, N., O'Toole, L., & Halpenny, A. (2017). *Introducing Bronfenbrenner*. London: Routledge.

Moen, P., Elder, G., & Luscher, K. (1995). *Examining lives in context: Perspectives on the ecology of human development*. New York: APA Science Net.

Rosa, E.M., & Tudge, J. (2013). Urie Bronfenbrenner's theory of human development: Its evolution from ecology to bioecology. *Journal of Family Theory & Review*, 5(4), 243–258. doi:10.1111/jftr.12022

Tudge, J. (2008). *The everyday lives of young children: Culture, class, and child rearing in diverse societies*. Cambridge: Cambridge University Press. doi:10.1017/CBO9780511499890

Tudge, J.R.H. (2017). Urie Bronfenbrenner. In Heather Montgomery (Ed.), *Oxford bibliographies on line: Childhood studies*. New York: Oxford University Press. doi:10.1093/OBO/9780199791231–0112

Tudge, J.R.H., Freitas, L.B.L., & Doucet, F. (2009). The transition to school: Reflections from a contextualist perspective. In H. Daniels, J. Porter, & H. Lauder (Eds.), *Educational theories, cultures and learning: Critical perspectives* (pp. 117–133). London: Routledge/Falmer.

Tudge, J.R., Merçon-Vargas, E.A., & Liang, Y. (2017). The importance of Urie Bronfenbrenner's bioecological theory for early childhood education. In *Theories of early childhood education* (pp. 67–79). London and New York: Routledge.

Tudge, J.R.H., Mokrova, I., Hatfield, B.E., & Karnik, R. (2009). The uses and misuses of Bronfenbrenner's bioecological theory of human development. *Journal of Family Theory and Review*, 1(4), 198–210.

Tudge, J.R.H., Payir, A., & Merçon-Vargas, E.A. (under review). Urie Bronfenbrenner's bioecological theory: Its development, core concepts, and critical issues. In K. Adamsons, A. Few-Demo, C. Proulx, & K. Roy (Eds.), *Sourcebook of family theories and methodologies*. New York: Springer.

Tudge, J.R.H., Payir, A., Merçon-Vargas, E.A., Cao, H., Liang, Y., Li, J., & O'Brien, L.T. (2016). Still misused after all these years? A re-evaluation of the uses of Bronfenbrenner's bioecological theory of human development. *Journal of Family Theory and Review*, 8, 427–445. doi:10.1111/jftr.12165

Tudge, J.R.H., Rosa, E.A., & Payir, A. (2018). Urie Bronfenbrenner's bioecological theory of human development. In M. Bornstein (Ed.), *The Sage encyclopedia of lifespan human development* (pp. 251–252). Thousand Oaks, CA: Sage.

5

THE IMPACT OF PSYCHOLOGICAL SCIENCES ON EARLY CHILDHOOD EDUCATION

Developmental Influences

David Elkind

While there are many theories in psychology that can apply to early childhood education, they are limited in scope. Sigmund Freud, Jean Piaget, and Erik Erikson have offered the only truly comprehensive theories of human development. These three theorists have, however, each addressed a different facet of the developing child. Freud focused upon the emotional, clinical side of development, Piaget centered upon the development of cognition and intelligence, while Erikson presented a view of the whole life cycle from a social and cultural interaction perspective. As such, each has contributed unique but complementary views of the child that led to unique but balanced early childhood education structures and strategies.

These three theories complement one another and, taken together, give us the most comprehensive depiction of early child development we have today. Some may object and argue that these writers are out of date and their work does not take account of the contemporary trends in research and theory. While that is true to some extent, it is also nonetheless true that we are still biological beings and no amount of innovation is going to change that. Human development is still limited by our biology and that is why the work of these innovators remains relevant today. Indeed, many of their conceptualizations are now part of our everyday discourse. The critical evidence of their lasting significance.

Within the constraints of a chapter, I will limit the presentation to only those parts of their work that pertains to early childhood – development and practice. A concluding summary will attempt to bring together some of their insights to provide a holistic conception of early childhood psychological development and practice.

Sigmund Freud (1856–1939)

Like Charles Darwin's theories, those of Freud brought about a paradigm shift (Kuhn, 1996) in our ways of thinking about our common humanity and our personal selves. Darwin deflated the biblical account of creation, while Freud challenged our belief that we are entirely rational beings. Significant parts of both theories are now widely accepted.

Freud, showing his talents early, like so many individuals of genius, was an outstanding student, and in his early teens he had a working knowledge of Greek, German, and Hebrew. He also taught himself the basics of English and Italian. At age 17, he entered the University of Vienna to study medicine. However, after attending a conference in Paris, he decided to become a natural scientist, to understand biological, and, eventually, psychological phenomena. He received his MD in 1881. During the 1885–1886 winter months, he went to study with famed neurologist Jean-Martin Charcot. Freud became fascinated with Charcot's work with psychological disturbances with no known neurological causes, termed *hysterias*. He was also intrigued by Charcot's use of hypnotism to cure these disorders. He returned to Vienna and opened a private practice as a neurologist.

Initially, Freud had few patients and worked together with a fellow neurologist Joseph Brewer. Brewer had a female patient with hysterical paralysis, which he cured with what he called the "talking cure," allowing the patient to talk freely about her condition. Freud had similar cases and he too used the talking cure. Together they published a book, *Studies in Hysteria* (Breuer, 1995/2000), documenting their cases, their methods, and the findings.

Freud continued seeing patients and soon acquired a reputation for his success with those with hysterical symptoms. With further clinical experience, he refined the talking cure, to the method of "Free Association," which encouraged the patient to say whatever came to mind. He also analyzed patients' dreams and kept a record of his own. In 1900 he published his now classic book, *The Interpretation of Dreams* (Freud, 1900/1913). As Freud's fame grew, he attracted a number of followers who eventually formed a Society for Psychoanalysis, what Freud called his method and theory of treating psychiatric patients.

Over the years, Freud's theories evolved, and he created four different models of the human personality: the dynamic, the developmental, the structural, and the functional. These models introduced the major concepts of psychoanalysis.

At the heart of Freud's theorizing was the importance of the sexual drive, which he believed was the cause of hysterical symptoms. His concept of sexuality, however, is much broader than the adult understanding of sex as physical attraction and sexual intercourse.

In the second essay in his book, *Three Essays on the Theory of Sexuality* (Freud, 1905/1962), Freud makes it clear that when he is speaking of infant sexuality, he is speaking of it in the broader sense of cyclical drives like hunger, thirst, and excretion. From this perspective, all of these cyclical drives follow a similar pattern of gradual build-up of pressure, followed by a sudden pleasurable release of tension. Freud did add that the sexual components of the hunger, thirst, and excretory function zones were more in the nature of fore-pleasure and not comparable to that associated with orgasm.

Freud's Four Models of the Personality

The Dynamic Model

This is the first model Freud introduced, and he used it to explain the appearance of hysterical symptoms. The sexual drive in all of its sequential manifestations creates pressure to be discharged but is kept from doing so by psychological repression. As a result, the built-up pressure finds its release in physical symptoms that have no physical determinants. The symptoms displace and redirect the pressure from the sexual drive to the body's musculature, which can produce paralysis of the arms, legs, or vocal cords.

The Developmental Model

In the *Three Essays* book, Freud also introduces a developmental model of the personality. He argues that the sexual drive, in the sense described above, moves from the oral, to the anal, and finally to the genital zones during the first five years of life. During the genital stage, the child is sexually attracted (Catheter) to the parent of the opposite sex. Freud labeled these attractions the *Oedipus Complex* for boys and the *Electra Complex* for girls. From the ages of 6 to 12, the sexual drive is in a state of dormancy. Freud termed this the *Latency Period*. In adolescence the sexual drive is reawakened but now the young person must give up the parent as a sexual love object and look elsewhere for a person to love.

The Structural Model

Freud's book, *The Ego and the Id* (Freud, 1923) describes his threefold structure of the mind. The *Id* represents the unconscious impulses, desires, and wishes of the individual (the seething cauldron of desire). The *Ego* emerges from the failure of the Id to realize its goal and serves as an adaptive mediator between the Id and external reality. For example, if the infant's cries for food are not immediately met, the ego emerges to delay gratification. At the age of about 4 or 5 the child identifies with the parent of the same sex as a means of establishing his or her sexual identity. Through this identification, the child internalizes the moral precepts of the parent, which becomes the *Superego*. The Superego is our conscience and the Ego must mediate between the demands of the Id and the constraints imposed by the Superego.

The Functional Model

Freud's functional model describes the mechanisms the Ego uses to mediate between the Id and the Superego. These are the ego defenses: *Repression, Projection, Reaction Formation*, and *Externalization*. Each of these mechanisms can distort reality in such a way that the Ego feels comfortable in satisfying the Id, even though the means to that end might not be entirely acceptable to the Superego. Yet the excessive use of such mechanisms may itself bring about mental pathology.

These four models give a fairly comprehensive picture of the dynamics of early childhood emotional life. For example, Freud's concept of infant cathexis to parents was translated by John Bowlby (1983) into the concept of attachment and all of the research on security of attachment stemming from that concept. Likewise, his concept of the Ego has now been translated and studied as the *Executive Function*. Moreover, his description of how over or under parental concerns with feeding and toilet training can lead to lasting personality orientations has also been studied. In addition, the concepts of the Oedipal and Electra Complexes help us to understand infantile dynamics. To illustrate, if a child's parents were to divorce during the early childhood period, the child may blame himself for having brought about the split thanks to his attachment to the opposite sex parent. Such children may become accident-prone. These are but a few examples of how Freud has helped us understand the emotional side of early childhood and the ways in which these ideas were integrated into the scientific knowledge upon which early childhood education was built.

Jean Piaget (1896–1980)

Just as Freud changed our understanding of human sexuality, Piaget broadened our understanding of human cognition's and intelligence's adaptive thought and action. Within psychology, the concept of intelligence has always been associated with individual differences, mental testing, and the IQ. For Piaget, in contrast, intelligence is a species characteristic that had to be looked at from social, biological, and logical perspectives.

Like Freud, Piaget showed his talents early. As an adolescent, he was mentored by an esteemed malacologist and published his first Paper on an Albino sparrow at the age of 13. He went on to receive his doctorate in biology at the University of Neuchatel. Not sure of his direction in life, he spent a year at Jung's psychiatric institute in Zurich. After this, still undecided on his vocation, he moved to Paris where he worked with Theodore Simon who, with Alfred Binet, created the first intelligence test (Binet, 1906/1916). While in Paris, Piaget translated into French, and tested out, some English items added to the Binet Scale by Lewis Terman (1916) at Stanford. What fascinated Piaget about Binet's tasks were not children's right answers but rather their wrong ones. Just as Freud (1951) recognized the importance of errors made in everyday life, Piaget appreciated that children's "wrong" answers gave significant insights into their unique modes of thinking.

From Paris, Piaget went on to take a professorship at the University of Geneva, where he remained for the rest of his career. Combining the interview skills he had learned in Zurich with the kinds of tasks introduced by Binet, he created the "semi-clinical interview," which he used to explore the development of children's thinking. In this way, he was able to combine his biological with his philosophical interests. He was most concerned with the epistemological question of "how we come to know the world" but by answers arrived at by empirical investigation rather than by armchair analysis.

As did Freud, Piaget successively created several models to describe his theory of intellectual development: the social, the biological, and the logical. The models were essentially three different ways of conceptualizing development. As he elaborated each new model, he integrated the previous one within it. Unfortunately, this made his later books increasingly more difficult to follow.

Piaget's Three Models of Mental Development

The Social Model

Piaget was influenced early on by the French sociologist Henri Bergson (Bergson, 1911), and did his first investigations from a sociological perspective. These were his studies of children's language (Piaget, 1955), as well as of their conceptions of physical causality (Piaget, 1930/1951), the world (Piaget, 1929/1951), of judgment and reasoning (Piaget, 1928/1951), and of their moral development (Piaget, 1950). In explaining the results of these investigations, he employed the term "egocentrism" to refer to the young child's inability to take another person's point of view. For example, he found that preschool children talk *at* one another rather than *to* one another. Egocentrism was used to explain his other findings as well.

Perhaps the most familiar task Piaget used to demonstrate egocentrism was the three mountains task. He employed clay models of three mountains placed at different positions on a table. The child's task was to draw a copy of the mountains as seen from the perspective of another child sitting opposite him. Again, young children's failure on this task was ascribed to the child's inability to put himself in another child's position.

The Biological Model

Piaget introduced his second model, the biological model, after his marriage and the birth of his three children. He engaged in a number of very innovative experiments with his infant daughters and son. These experiments changed his focus from the social to the biological and Piaget looked at the behavioral mechanisms by which infants progressively construct their understanding of reality (Piaget, 1954), their intelligence (Piaget, 1952a), and their dreams (Piaget, 1951). The biological concepts that Piaget employed were not those of stimulus and response, as employed by American psychologists. Rather, he used the broader concepts of adaptation: those of assimilation, accommodation, and equilibration. This is essentially a dialectic model in which the infant progressively accommodates to, and then assimilates contradictory experiences before uniting them in a higher order equilibrium. For example, the infant first accommodates his or her mouth to the nipple, then assimilates the milk as nourishment and then coordinates the two into a single higher order equilibrated ingestion activity.

Along with the biological model, Piaget also formally introduced an epistemological theory that was already implicit in his earlier work. In his view, reality is always a construction. It is neither a simple copy of external reality nor a simple projection of internal constructions. An analogy may help to concretize this view.

Traditional association theories posit that all knowledge comes from experience, with the mind operating as a camera, progressively taking photographs of the environment as the child matures. Alternatively, the Nativist theory, first introduced by Plato, assumes that all knowledge is already in the mind and can be discovered by dialectical, logical discussion. In contrast, Piaget argues that what we know always reflects both input from the outside, accommodation, and input from the inside, assimilation. The analogy here would be the creative artist in any discipline. A Picasso painting, for example, is always recognizable as a Picasso, yet each painting also reflects his experiences of the real world. It is both individual and social, in that it reflects his uniqueness but also is able to be appreciated by others. It is a construction, which cannot be reduced to nature (Picasso) or nurture (the environment). Even our perception is, in part at least, a mental construction and children literally see the world differently as they mature (Elkind et al., 1963).

The Logical Model

In the late 1940s Piaget introduced a logical model to describe the changes in intelligence, which he documented with a large number of investigations centered on the issues of "conservation"; how we come to understand that things remain the same despite a change in appearance. He argued that changes in a child's ability to conserve concepts such as number, space, and time are a result of attaining new reasoning abilities. He used this model to give a more complete picture of mental development than he had with the previous models. He divided intellectual development into four stages distinguished by the logical structures that characterized each stage.

The first stage in the development of intelligence, Piaget calls the *sensory-motor stage* or period (usually birth–2 years). This stage is concerned with the evolution of those abilities necessary to mentally "construct" permanent objects. Although it seems to us that we have known from birth that objects outside of ourselves continue to exist when they are no longer present to our senses, this turns out not to be the case. For the infant, out of sight is literally out of mind. If an infant is playing with a rattle and it falls from the crib, she is not upset and finds some other way to amuse herself. After the first year, however, the infant has been able to construct a mental image of the object by coordinating all of her sense experiences with it. By that point, if the same rattle disappears, the infant becomes upset and searches for it.

The second stage in the intellectual development is what Piaget calls the *preoperational stage* (usually 2–6 years). At this stage the child gradually acquires the symbolic function – those abilities dealing with how we represent things, words, and symbols. The presence of these new abilities is shown by the gradual acquisition of language, the first indications of dreams and night terrors, and by the advent of symbolic play (two sticks at right angles are an airplane), and the first attempts at drawing and graphic representation.

The preoperational stage is also characterized by what Piaget called *transductive* thinking. With deductive or inductive thinking, one reasons from class to individual or from individual to class (e.g., from fruit to apple or from apple to fruit). But transductive reasoning

moves from object to function at the same concrete level. Such thinking is apparent when a preschooler is asked to define the word "bike" as "something you ride" or an "apple" as "something to eat." In contrast, an older child would say that a bike is a "vehicle" or a "means of transportation" and an apple is a "fruit." Children at this stage therefore reason only at a concrete level. It is because preschoolers think in this way that they are not able to nest one class inside of another. A preschooler for example may know the number of boys and girls in his class but is baffled by the question, "Are there more boys or more children in the class?" The child at this stage is not yet able to understand that one can be both a boy and a child at the same time. It is difficult for children to reason beyond that which they can perceive directly. They have difficulty making reasoned decisions regarding subordinate classes (boy) and superordinate classes (children). This fact is exemplified by the example of a 4-year-old who encounters his teacher in a store, and is upset and confused because he cannot believe someone can be a person and a teacher at the same time! It is only at the next stage that children are able to understand that you can be both an individual and a member of a class of individuals.

The third stage in the intellectual development is what Piaget refers to as the *concrete operational stage* (usually 7–11 years). By the age of 7, most children have acquired what Piaget calls *concrete operations*, internalized actions that permit the child to do in his head what, before, he would have had to accomplish through real actions. This is what the ancients called the "Age of Reason": the age at which they could engage in deductive thinking.

Piaget's later investigations focused on how children use their new reasoning ability to distinguish between the permanent and the transient, and between appearance and reality. During this stage he and his colleagues attacked the child's developmentally attained understanding of the basic philosophical concepts such as Space (Piaget, 1967), Time (Piaget, 1969), and Number (Piaget, 1952b).

One of the experiments Piaget employed to study the development of the concept of number was the "six pennies task." The child was shown a row of six pennies and asked to take the same number from a pile of pennies. Children of 3 or 4 either made a line of more pennies, but of the same length, or of more pennies by extending the line using the same spacing as in the original. Children of 4 or 5 matched the pennies in the row. When, however, the experiment spread the pennies in one row further apart than in the other row, the child usually said the longer row had more. It was only at the age of 6 or 7 that most children said the long and short rows both had the same number.

Piaget argued that the youngest children judged number by a single dimension, length, or density. Somewhat older children could make the one-to-one correspondence but could not overcome the perceptual impression that the longer line had more. It was only the older children who said that no pennies were added or removed so they remained the same. Yet the younger children also knew that no pennies had been added or taken away from the two rows of unequal length. To solve the problem the child had to reason that what the row gained in length it lost in density. However, that reasoning was unconscious so the child could not verbalize it. As Freud pointed out, much of our emotional thinking is unconscious and Piaget argued that much of our reasoning is unconscious as well. We become aware of our reasoning only on reflection, not while we are in the process of thinking.

It is only with attainment of concrete operations that children do math and comprehend phonics. In both skills the child has to grasp that one thing can be two things at the same time (that number is both a place and a quantity and that letters can represent more than one sound and that one sound can be signaled by more than one letter).

As shown above, Piaget's theory has provided valuable insights for both parenting and educational instruction. With regard to parenting, the concept of concrete operations explains why it makes little sense to reason with a preschooler who does not yet have mental tools. And it also explains why it is a mistake for teachers to try to teach phonics or arithmetic operations to children who have not attained the age of reason. Piaget's studies and scientific understanding of intellectual development support the use of manipulatives in preparation for mathematical instruction inasmuch as operations are internalized actions. Most importantly, Piaget's work makes clear that young children's thinking is not wrong, it is age-appropriate, just qualitatively different from older children's or adult thinking. We have learned from Piaget that curriculum content, teaching strategies, and learning materials must be in accord with young children's intellectual capacities. These are but a few of the valuable practical implications of Piaget's research and theory for both parenting and education.

Erik Erikson (1902–1994)

In many respects, Erik Erikson differs from both Freud and Piaget. Freud and Piaget both spent most of their lives in the same place whereas Erikson, who was born in Denmark, spent his youth traveling in Europe before settling in Vienna. He left Europe for the United States in 1936 where he taught at several major universities and did clinical work at Austen Riggs, a residential treatment center in Stockbridge MA. He also differed from Piaget and Freud in that although he held professorships at Harvard, Yale, and Berkeley, he never received a college degree.

The explanation for the latter fact comes from his training as a child analyst; it was a new field introduced by Anna Freud and Dorothy Burlingham. Erikson was analyzed by Anna Freud and became one of the few child psychoanalysts in the world. Erikson's remarkable clinical skills made him in demand at major universities in the growing academic departments of clinical psychology. He ended his career having an Endowed Chair at Harvard.

Although Erikson expanded upon Freudian theory, he never broke with Sigmund or Anna Freud, unlike Jung and Adler. Rather, he saw himself as extending Freudian theory to the social domain and to the whole life cycle. His first major work was *Childhood and Society* (Erikson, 1950). He went on to write books on Gandhi (Erikson, 1969) and Martin Luther (Erikson, 1950; 1958) in which he used his concepts of *Identity* and *Identity Crisis* to explain, in part at least, the achievements of these two extraordinary men.

In *Childhood and Society*, Erikson describes the eight stages of human development for which he is best known. In his elaboration of those stages, he built upon Freud's sexual stages but added to them concepts of health and pathology across the whole life cycle. In Erikson's view we are endowed with eight social "senses," which, while all present at birth and throughout life, have major critical periods during the life cycle. It is during these periods that each sense attains lasting strength or weakness. However, Erikson also believed that we are resilient and a bad outcome at one stage can be remedied at a later stage but with greater effort than would have been required earlier.

Erikson's Stages

Infancy: Trust vs. Mistrust

According to Erikson, in the first social sense (corresponding to Freud's anal stage) the child's experiences are that of trust; the sense that the world is a safe place, that one's needs will be met. To a large extent the infant's sense of trust in the world, others, and herself depends upon the quality of care giving she receives during this age period. The infant whose needs are met when they arise, whose discomforts are quickly removed, and who is cuddled, played with, talked with, and sung to, develops a sense of the environment as a safe place and of people as helpful and dependable. When, however, care is inconsistent, inadequate, and/or rejecting, it fosters basic mistrust of others and the world.

Toddlerhood: Autonomy vs. Shame and Doubt

The period that Freud labeled the "anal" period is that of the second and third years of life, when among other things the toddler attains bowel and bladder control. Erikson says that the sense of autonomy, or personal control of the self, is a critical attainment of this period. At this stage the child can walk and climb, open and close, push and pull, hold and let go. The child now wants to do everything by himself, including toilet training. If parents recognize the child's need to do what he is capable of doing at his own pace and time, then the child attains a send of being in control of his muscles, impulses, self, and the environment. Obviously, such control is far from perfect, and autonomy is also fostered when parents take accidents as part of the maturing process. If, however, parents are impatient for the child to be toilet trained and do for him what they are capable of doing for themselves, they discourage autonomy. Instead, the child develops a lasting sense of shame and doubt about his own abilities to control self and environment.

Early Childhood: Initiative vs. Guilt

As the child becomes more proficient physically, mentally, and socially, she also has an increased sense of initiative, which is often expressed as curiosity, question asking, and a willingness to take on activities such as household chores. Parents encourage this sense of initiative by supporting children's curiosity, answering children's questions and letting them engage in age-appropriate household tasks, such as picking up their toys, putting soiled clothes in a hamper, etc. If, on the other hand, parents are impatient of the children's curious dawdling, bothered by their questions and their desire to be of service, they encourage a sense of guilt, of always doing something bad or hurtful to others. As children develop, this may be reflected in a kind of accident proneness, of hurting themselves as punishment for their supposed wrongdoing, interfering with their performance both at home and at school.

Erikson appreciates that a sense of mistrust, shame, doubt, and guilt can be overcome by later good experiences, but overcoming these negative attributes is more difficult the later in life the more positive experiences occur. Likewise, the sense of trust, autonomy, and initiative can also be undone by later bad experiences, but these are less threatened the more established they become.

Erikson's work has considerable impact both on childrearing and education. His concepts of trust and mistrust have reinforced the importance of physical contact, a comforting environment, and regularity of routine in infant care giving. The concepts of autonomy highlight the importance of following the child's lead when it comes to toilet training and not to be bothered by the 2-year-old's negativity in his struggle to attain autonomy. And the concept of initiative has added strength to the idea that preschoolers need to be allowed to explore and create within age-appropriate limits.

Summary and Conclusion

In this chapter, three developmental scientific views of children were presented. As was stated earlier, these three views of the child, while offering unique perspectives, are at the same time complementary to each other. As such, they have contributed to balancing early childhood knowledge and practice so that all facets of children's developmental needs are met. One way to show how the theories described above complement one another is to illustrate this at each stage of early childhood development.

Infancy

For Freud, infancy begins with the oral stage and the infant's needs to be met. For Piaget, it is the period when the infant constructs a permanent object that can be expected to continue to meet the child's needs. One could put this differently by saying that the infant's Ego progressively constructs a permanent object to meet the Id's needs. At the same time, the permanent object also provides the infant with a sense of trust. The sense of *objecthood* integrates the attainments made during the infancy period

Toddlerhood

Around the second year of life Freud posits that the anus becomes the focus of pleasure and distress. For Piaget, this is a time when the child begins to acquire the symbolic system of language, dreams, and other symbols. One might argue that the development of language is one of the means the Ego uses to bring impulses under control. Indeed, the toddler can now indicate when she needs the bathroom. For Erikson, the symbolic system provided a new means for the ego to seek autonomy by asserting the self and saying "no." A sense of *Personhood* brings together the overall attainment of this stage.

Early Childhood

In the Freudian scheme, this is the age when the child identifies with the parent of the opposite sex and begins to internalize the Superego. At this stage, the child begins to construct and accept a sense of sexual identity. For Piaget, this is the stage of preoperations when the child engages in magical thinking, and believes in fairy tales, Santa Claus, and the Tooth Fairy. One might argue that magical thinking plays a role in the child's identification with the parent who may be seen in Godlike terms. Indeed, young children believe that everything has a purpose and that parents know that purpose. In the Eriksonian scheme, this is the period when the child gains a sense of initiative. When the child identifies with the parent, he wants to emulate the parent's ability to initiate activities and be in control of world. This sense of initiative may be helped or hindered by the parent's willingness to let the child engage in meaningful tasks. The sense *of selfhood* integrates this new sense of being in touch with the world as well as with others.

This chapter offers only a brief illustration of how these three scientific theories dovetail to give a more complete representation of early childhood development. Freud's scientific thinking provided insights into the conscious and unconscious drives that motivate children's behaviors. His thoughts and ideas ultimately led to ideas that are prevalent in early childhood such as the "whole child," "play-based pedagogy", and "creativity" as the basis for early childhood education. Piaget's view was that children are unable to commence particular undertakings until they are psychologically mature enough to do so. The implications of this for early childhood education are that teaching strategies and curriculum must be adapted according to the developmental level of the child. Importantly, this includes early childhood education teachers' awareness of how children's psychological development affects their understanding of self and their behavioral interpretation of social and emotional appropriateness. Finally, Erikson provided the framework for understanding that cognitive development and social/emotional development were interconnected. In early childhood education, teachers play a significant role in supporting young children's psychosocial growth and development.

Together, Freud, Piaget, and Erikson have provided the scientific framework for understanding children's development from a comprehensive perspective. Throughout the years, early childhood education has benefited from this perspective in being able to provide theoretically informed teaching and curriculum.

References

Bergson, H. (1911). *Creative evolution*. New York: MacMillan.

Binet, S.a.S., Th. (1906/1916). *The development of intelligence in children*. Baltimore, MD: Williams and Wilkins.

Bowlby, J. (1983). *Attachment*. New York: Basic Books.

Breuer, J.F., Sigmund (1995/2000). *Studies in hysteria*. New York: Basic Books.

Elkind, D., *et al.* (1963). Effects of perceptual training at three age levels. *Science*, 137, 3532–3537.

Erikson, E. (1950). *Childhood and society*. New York: Norton.

Erikson, E. (1958). *Young man Luther: A study of psychoanalysis and history*. New York: Norton.

Erikson, E. (1969). *Gandhi's truth: On the origins of militant non-violence*. New York: Norton.

Freud, S. (1900/1913). *The interpretation of dreams*. New York: MacMillan & Co.

Freud, S. (1905/1962). *Three Essays on the Theory of Sexuality*. Trans. James Strachey. New York: Basic Books.

Freud, S. (1923). *The ego and the id*. New York: Norton.

Freud, S. (1951). *The psychopathology of everyday life*. New York: Mentor.

Kuhn, T. (1996). *The structure of scientific revolutions*. Chicago: University of Chicago Press.

Piaget, J. (1928/1951). *Judgement and reasoning in the child*. London: Routledge & Kegan Paul.

Piaget, J. (1929/1951). *The child's conception of the world*. London: Routledge & Kegan Paul.

Piaget, J. (1930/1951). *The child's conception of physical causality*. London: Routledge & Kegan Paul.

Piaget, J. (1950). *The moral judgement of the child*. London: Routledge & Kegan Paul.

Piaget, J. (1951). *Play, dreams and imitation in childhood*. Melbourne: Heinemann.

Piaget, J. (1952a). *The origins of intelligence in the child*. New York: International Universities Press.

Piaget, J. (1952b). *The child's conception of number*. London: Routledge and Kegan Paul.

Piaget, J. (1954). *The construction of reality in the child*. New York: Basic Books.

Piaget, J. (1955). *The language and thought of the child*. New York: Meridian.

Piaget, J. (1967). *The child's conception of space*. London: Routledge, Kegan Paul.

Piaget, J. (1969). *The child's conception of time*. London: Routledge Kegan Paul.

Terman, L. (1916). *The measurement of intelligence*. Stanford, CA: Stanford University Press.

6

THE IMPACT OF PSYCHOLOGICAL SCIENCES ON EARLY CHILDHOOD EDUCATION

Child's Mind in Society – A Vygotskian Approach to Early Childhood Education

Elena Bodrova and Deborah J. Leong

This approach is known in the West under many different names: it is referred to as "socio-cultural," "social-constructivist," and even (and quite confusingly so) "social learning." All these definitions underscore one of the major distinguishing features of the Vygotskian approach – its emphasis on the role of social interactions in the processes of children's learning and development. Although these terms are useful when contrasting Vygotskian theory with the nativist theories of development or the behaviorist theories of learning, the use of such terms can sometimes lead to misunderstanding or oversimplification of the key Vygotskian ideas.

Vygotsky and his close colleagues used a different term – Cultural-Historical – to describe their approach and this term is now used by the scholars who can trace their philosophical lineage to Vygotsky's students or the students of Vygotsky's students. To better understand Vygotskian views on child development and education it is important to "unpack" the concept of the Cultural-Historical approach and to have a more nuanced understanding of its meaning.

We would like to point out one thing in particular: the omission of the "history" part from the description of the Vygotskian approach undermines the developmental focus of this approach. This focus is especially relevant when we attempt to apply Vygotskian theory to early childhood education since the same social interactions and cultural-specific factors might play a different role depending on where a child is in his or her individual history of development and also depending on when and why these interactions and these factors appeared in the history of humankind.

It is also worth noting that Vygotskians do not see the Early Childhood classroom as merely a place to apply their theories of learning and development. Instead, they see classrooms as "laboratories" to study child development as it is being shaped by the social context. This approach extends to the programs designed to serve children with special needs as well as the programs designed to replace parental care such as orphanages and boarding schools. Integrating research done in all these varying contexts allowed Vygotsky and his students to generate a rich theory describing child development in a social context.

Cultural-Historical Theory: Defining the Terms

Lev Vygotsky (1896–1934), along with Alexander Luria, Alexei Leont'ev, and the others collectively known as the "Vygotsky circle," embarked on an ambitious project of developing a "new psychology" that would "sublate" (in a Hegelian sense) the psychological theories most influential in the first quarter of the twentieth century: the objectivist psychology that focused on observable behaviors common to humans and animals alike and the subjectivist psychology concerned with the uniquely human experiences. This project was not destined to be completed due to many factors including Vygotsky's untimely death as well as the changes in the policies of the Soviet state in the areas of education and social sciences that caused a halt in further work on the theory. Through the efforts of Vygotsky's students (who were forced to work in a "clandestine" manner until the political "thaw" of the 1960s) many revolutionary ideas of their mentor have been further developed and applied in various fields ranging from literary criticism to neuroscience to special education. This work led to the formation of the body of knowledge defined as Cultural-Historical theory.

The term *culture*, as in the Cultural-Historical approach, was used by Vygotsky in two major ways: in the discussion of specific socio-cultural environments that shape children's learning and development and also in the study of cultural tools such as signs and symbols and their role in human development (Vygotsky, 1997). It seems that this first aspect of culture received most attention in the work of post-Vygotskians in the West while in-depth studies of children's appropriation of specific cultural tools and of the resulting changes in their psychological processes remain relatively few. At the same time, it is this part of Vygotsky's legacy that makes it relevant for solving today's challenges of education in general and especially of Early Childhood Education.

The term *history* in this approach is also used in a very specific context: Vygotsky proposed that to truly understand psychological processes one needs to study these processes as they change and develop, hence the focus on the history. Vygotsky argued that fully developed psychological processes are hard to study due to the fact that they usually exist in an internalized and "folded" form where many of the component processes are not easily accessible. Therefore, the methodology commonly employed by developmental psychologists such as longitudinal or cross-sectional design can only describe an outcome but not the process itself. At the same time, a process undergoing development still has an extensive external component accessible to observation which may provide researchers with an insight into the nature of this particular process (Vygotsky, 1978). The word *historical* in the Cultural-Historical approach applies to the study of the development of psychological processes as they unfold through an individual person's history or *ontogeny* and the history of humankind or *phylogeny*. The latter focus explains Vygotsky's special interest in the development of children at the age when these processes are still taking shape.

The Concepts of Cultural Tools and Higher Mental Functions

Cultural tools is one of the central concepts in Vygotsky's theory. While defining humans as "tool-making animals" was quite common in Vygotsky's time, this definition was only applied to physical (mechanical) tools. Vygotsky extended this idea to a new kind of tools – cultural tools – positing that it is this particular kind of tools that sets humans apart from other animals. Cultural tools – from simple smoke signals to elaborate writing systems – allowed humans take control over their own psychological processes:

> The person, using the power of things or stimuli, controls his own behavior through them, grouping them, putting them together, sorting them. In other words, the great uniqueness of the will consists of man having no power over his own behavior other than the power that things have over his behavior. But man subjects to himself the

> power of things over behavior, makes them serve his own purposes and controls that power as he wants. He changes the environment with the external activity and in this way affects his own behavior, subjecting it to his own authority.
>
> *(Vygotsky, 1997, p. 212)*

Thus, Vygotsky argued, similar to the way physical tools extend one's physical abilities by acting as an extension of the body, cultural tools extend one's mental abilities, acting as an extension of the mind. The use of cultural tools made it possible for humans to engage in increasingly more complex behaviors, thus starting humankind on a path of cultural evolution that had largely replaced the biological one. In their discussion of cultural tools, Vygotsky scholars often refer to them as *mental tools* or *tools of the mind*, so in this chapter we will be using these terms interchangeably.

In his own writings, Vygotsky focused primarily on language-based mental tools, from gestures to oral speech to written speech, and the ways the use of these tools transforms humans' thinking. This exclusive focus on language became one of the most frequent criticisms of Vygotsky's theory and possibly one of the barriers to its broader application in educational practice with young children who are still in the process of mastering language skills.

Subsequent generations of post-Vygotskians applied the idea of cultural tools to other areas of child development and demonstrated how mastery of a variety of tools, including non-linguistic ones, can transform developing psychological processes including perception, attention, memory, and problem solving. Examples of non-linguistic mental tools include *sensory standards*, which reflect "socially elaborated patterns of sensory characteristics of objects" (Venger, 1988, p. 148) such as colors, shapes, musical tones, and so on. Another example of children's use of non-linguistic mental tools is described in the work of a Vygotskian scholar, Piotr Gal'perin (Gal'perin & Georgiev, 1960) where kindergarten-age children were taught the concept of number through a series of measurement activities using specially designed manipulatives.

Children learn mental tools from adult members of their society or from their older peers, described by Vygotsky as "more knowledgeable others." Therefore, for Vygotsky, one of the major goals of education, formal as well as informal, is to help children acquire the tools of their culture (Karpov, 2005). This places the emphasis on learning how to use the tool to solve problems, not just on the tool itself. Teaching children how to use mental tools results in children mastering their own behavior, gaining independence, and reaching a higher level of development. For Vygotsky, acquiring and practicing an increasing number of mental tools not only transforms children's external behaviors but shapes their very minds, leading to the emergence of a new category of mental functions – *higher mental functions*.

Vygotsky's theory of higher and lower mental functions was his answer to the "nature vs. nurture" debate that had started long before his time and in some form is still going on today. Vygotsky (1997) describes *lower mental functions* as common to humans and higher animals, innate, and depending primarily on maturation to develop. At birth, lower mental functions in humans are culturally independent; however, they may be later transformed and re-structured as a result of the development of higher mental functions and their further development may take a culturally specific form. As children start utilizing higher mental functions more frequently, their lower mental functions do not disappear completely, but get transformed and integrated with the higher mental functions:

> In the thinking of the adolescent, not only completely new complex synthetic forms that the three-year-old does not know arise, but even those elementary, primitive forms that the child of three has acquired are restructured on new bases during the transitional age.
>
> *(Vygotsky, 1998, p. 37)*

Examples of lower mental functions include sensations, spontaneous attention, associative memory, and sensorimotor intelligence. Sensation refers to using any of the senses and is determined by the anatomy and physiology of a sensory system in a particular species; reactive attention refers to attention that is drawn to strong or novel environmental stimuli; associative memory is the ability to connect two stimuli together in memory after repeated presentation of the two together; and sensorimotor intelligence describes problem solving via trial and error, using physical or motor manipulations.

Unique to humans, *higher mental functions* are cognitive processes children acquire through learning and using mental tools specific to their culture (Vygotsky, 1997). For example, learning and using color words would result in words mediating what children notice and cause children to be able to differentiate between more or fewer shades depending on the size of their color vocabulary. Thus, for Vygotsky, *mediated perception* would be an example of a higher mental function that transforms the way humans see colors without making changes to anatomy or physiology of their visual apparatus. In a similar way, children's mastery of mental tools transforms other lower mental functions, leading to the development of focused attention, deliberate memory, and logical thinking.

According to Vygotsky, higher mental functions in their mature form are deliberate, mediated, and internalized (Vygotsky, 1997). By characterizing higher mental functions as *deliberate*, Vygotsky implies that they are intentionally controlled by humans and not involuntary reactions elicited by an environmental stimulus: the behaviors guided by higher mental functions can be focused on specific aspects of the environment, such as ideas, perceptions, and images, while ignoring other inputs. In Vygotsky's own words, (Vygotsky, 1978) as children develop higher mental functions they begin to "master [their] attention" (p. 26) and "control [their] own behavior" (p. 28).

These deliberate behaviors become possible because they are not elicited by environmental stimuli in an immediate and direct fashion but instead are *mediated* by the use of tools. In his earlier works, Vygotsky (1994) writes about tools as specifically created intermediaries between environmental stimuli and human responses thus presenting an alternative to a S–R behaviorist model popular in his time. By the time higher mental functions are fully developed, most of the tools used are not external but internal (such as categories) and so are the processes involved in using these tools. Vygotsky describes this process as *internalization* emphasizing that when external behaviors "turn inward," they maintain the same focus, and function as their external precursors (Vygotsky, 1978, p. 57). For example, counting that starts in young children as a physical action involving the use of tangible objects (e.g., counters or fingers) eventually evolves into a cognitive process that does not depend on external actions or objects.

Vygotsky describes the development of higher mental functions as a gradual process involving the transition from inter-individual ("intermental") or shared to individual ("intramental"). The idea of "shared" or "intermental" is one of the most commonly misunderstood Vygotskian concepts. It is often interpreted as children's individual mental processes being affected by the social interactions this child engages in (somewhat similar to Piaget's view of how children are forced to resolve cognitive disequilibrium that resulted from their interaction with peers). Vygotsky, in fact, made a much stronger claim arguing that higher mental functions evolve out of social interactions continuing to bear some characteristics of these social interactions even in their developed form.

Higher mental functions begin as shared: they are co-constructed by the child in interaction with another person or persons. For young children, most of their higher mental functions still exist only in their shared form as these functions are still undergoing the process of co-construction (Vygotsky, 1997).

> Every function in the cultural development of the child appears on the stage twice, in two planes, first, the social, then the psychological, first between people as an intermental category, then within the child as intramental category. This pertains equally to voluntary attention, to logical memory, to the formation of concepts, and to the development of will.
>
> *(p. 106)*

We can illustrate the Vygotskian idea of gradual transition of a higher mental function from inter- to intramental using an example of how children develop the ability to regulate their actions. At first, to perform a voluntary (as opposed to spontaneous) action, a child needs to be regulated by someone else: a parent or a teacher. At this – intermental – stage, the initiation and the execution of the action is *shared* between two individuals: "I order, you execute." At the next stage, the child can both regulate and execute an action by issuing commands to him/herself and then executing these self-commands (picture a 2-year-old reaching for a hot stove while saying "No!" to herself). This stage is a transitional one: the child is learning to use the same tool – language – as the adult was using in the past but the tool itself is not yet internalized and does not always work. Finally, the external speech turns into an inaudible mental command and the voluntary action gets formed. For Vygotsky, the transition from inter- to intramental state of a mental function results in the formation of a new structural-functional system in the brain: "two points in the brain which are excited from outside have the tendency to work in a unified system and turn into an intra-cortical point" (Vygotsky, 1997, p. 106).

Vygotsky's view of child development differs therefore from other developmental theories because while acknowledging other people's influences on a person's individual development, these theories hold that all mental processes still ultimately reside in this individual's mind. In contrast, for Vygotsky, not only what a child knows but also how this child thinks, remembers, or attends is shaped by the child's current and prior interactions with parents, teachers, and peers.

Vygotsky's View of Child Development

As we stated above, the nature of the cultural tools acquired, as well as the nature of higher mental functions developed as the outcome of their acquisition, are both determined by the specific interactions that occur between children and their social environment. Vygotsky called these interactions the "social situation of development," which he considered to be the "basic source" of development. The social situation of development drives the transition from one stage in child development to the next. The social situation of development (Vygotsky, 1998)

> represents the initial moment for all dynamic changes that occur in development during the given period. It determines wholly and completely the forms and the path along

> which the child will acquire ever newer personality characteristics, drawing them from social reality as from the basic source of development, the path along which the social becomes the individual.
>
> *(p. 198)*

The idea of the social situation of development lies in the foundation of Vygotsky's view of child development including the development during early years. Early childhood for Vygotsky consists of three distinct stages or "age periods," each built on the foundation of the previous one (Karpov, 2005). Infancy describes the period from birth to approximately 12 months of age; toddlerhood (or "early age" in Vygotskian terms) lasts from 12 to 36 months, and preschool age lasts until the time of school entry. It should be noted that, in Russia, children start school at age 7, so "preschool age" for Vygotsky extends into the years that in other countries are considered "school age." Vygotsky's "age periods" are different from the views of stages shared by Jean Piaget and other developmental theorists: these "age periods" are as much social formations as they are biological constructs.

While Vygotsky had not completed his theory of child development in his lifetime, his writings indicate that he viewed child development as a series of stable periods alternating with what he called "critical periods" (Vygotsky, 1998). During these critical periods, the entire system of mental functions undergoes major restructuring that results in the emergence of cognitive and social-emotional "neo-formations" or developmental accomplishments. During stable periods no neo-formations appear and children still continue to develop their existing competencies.

One of the observable signs of children going through critical periods (or "crises" in Vygotsky's words) is that their behavior changes in a way often perceived by adults as negative: children who used to be easy and compliant start "misbehaving." Vygotsky explained these sudden changes as the indication that a child's new emerging needs come into conflict with the constraints imposed on them by their current social situation of development. Overcoming this conflict propels the child to the next developmental level. Vygotsky identified the typical ages associated with these crises in early childhood as 1, 3, and 7 years. These turning points signify the transitions from infancy to toddlerhood, from toddlerhood to preschool age, and from preschool age to school age.

Vygotsky's original concept of stable and critical periods was refined and expanded in the work of post-Vygotskians to form a theory that contains not only well-defined developmental stages but also the explanation of the mechanisms underlying children's transitions from one stage to the next (Elkonin, 1972; Karpov, 2005). One of the major contributions made by the post-Vygotskians to Vygotsky's theory of child development is the idea of *leading activity* – the idea that further elaborated Vygotsky's concept of the social situation of development.

Leading activity is defined as a type of interaction between children and their social environment that is the most beneficial for their development at this particular age period:

> Some types of activity are the leading ones at a given stage and are of greatest significance for the individual's subsequent development, and others are less important. We can say accordingly, that each stage of psychic development is characterized by a definite relation of the child to reality that is the leading one at that stage and by a definite, leading type of activity.
>
> *(Leont'ev, 1981, p. 395)*

Children's engagement in leading activity leads to the emergence of the neo-formations of this age period (Elkonin, 1972; Leont'ev, 1981). Neo-formations, in turn, are defined as competencies and skills that are not only new to a specific age period, but are also critical for the child's ability to engage in the leading activity of the following age period (Karpov, 2005). For example, the ability to use objects in a pretend way that signals the emergence of symbolic thinking and appears by the end of toddlerhood is necessary for the development of make-believe play, which is a leading activity of the preschool age.

Vygotskian scholars continue to study developmental and socio-cultural factors involved in children's engagement in leading activities and in the emergence of neo-formations. For example, the work of Maya Lisina and her students led to the identification of different formats of adult–child and child–child interactions at different times during early childhood and the effect of these interactions on children's development of neo-formations in families and in institutions (Lisina, 1985). Elena Kravtsova – another Vygotskian scholar and a granddaughter of Lev Vygotsky – focuses on the role of critical and stable periods in the development of neo-formations as well as the nature of adult–child interactions influencing child engagement in leading activities over the course of early childhood (Kravtsova, 2006).

Vygotskian Views of Play and Its Role in Child Development

When describing the Vygotskian approach to play it is important to emphasize that in their definition of play as the leading activity, Vygotskians focus on a specific kind of play: what is described in the literature as pretend, socio-dramatic, or make-believe play. It means that this approach may not apply to activities such as movement activities, games, object manipulations, and explorations that were (and still are) referred to as "play" by most educators and non-educators alike. In addition, the characteristics of play in the works of Vygotsky and his students imply that they describe what was later called a "fully developed" form of play (Elkonin, 2005) and not play in its early stages as it exists in toddlers and younger preschoolers.

Vygotsky described this "fully developed" play – play as a leading activity – as having three major features: children create an imaginary situation, take on and act out roles, and follow a set of rules determined by specific roles (Vygotsky, 1978). Each of these features plays an important role in children's development, shaping their higher mental functions. Role-playing in an imaginary situation requires children to engage in two types of actions: external and internal. In play, these internal actions, "operations with meanings" (Vygotsky, 1967 p. 15), are still dependent on the external manipulations of the objects. However, the very emergence of the internal actions signals the beginning of a child's transition from the earlier forms of thinking – sensory-motor and visual-representational – to more advanced symbolic thought:

> A child learns to consciously recognize his own actions and becomes aware that every object has a meaning. From the point of view of development, the fact of creating an imaginary situation can be regarded as a means of developing abstract thought.
>
> *(Vygotsky 1967, p. 17)*

Make-believe play therefore contributes to the development of two higher mental functions: thinking and imagination. Contrary to a popular belief that play is driven by children's imagination, Vygotsky considered imagination an outgrowth of play that emerges when children no longer need toys and props that serve as physical "pivots" helping them assign a new meaning to an existing object:

> Imagination is a new formation that is not present in the consciousness of the very young child, is totally absent in animals, and represents a specifically human form of conscious activity. Like all functions of consciousness, it originally arises from action. The old adage that children's play is imagination in action can be reversed: we can say that imagination in adolescents and schoolchildren is play without action.
>
> *(Vygotsky, 1967, p. 8)*

According to Vygotsky, the role of play in the development of higher mental functions has yet another aspect: it promotes children's intentional, deliberate behaviors. In this respect, the Vygotskian view of play differs from many other theories that see play as an activity where children are totally free of any constraints. Moreover, Vygotsky's student Daniil Elkonin, who extended these ideas into a Cultural-Historical play theory, called play "the school of deliberate behavior" (Elkonin, 1978, p. 287).

The need to act in a deliberate fashion is dictated by the inherent relationship that exists between the roles children play, the pretend props they use, and the rules they need to follow when playing these roles and using these props. For young children, play becomes the first activity where they are driven not by the need for instant gratification, prevalent at this age, but instead by the need to suppress their immediate impulses, their "reactive" behavior:

> The role the child plays, and her relationship to the object if the object has changed its meaning, will always stem from the rules, i.e., the imaginary situation will always contain rules. In play the child is free. But this is an illusory freedom.
>
> *(Vygotsky 1967, p. 10)*

These contributions of play in the development of higher mental functions led Vygotsky (1967) to define play as "not the predominant feature of childhood but is a leading factor in development" (p. 6). While for Vygotsky the term "leading activity" was more of a metaphor than a theoretical construct, Alexei Leont'ev (1981) and Daniel Elkonin (1977) further elaborated on this idea, specifying the distinct features of the leading activities characteristic of children at different ages. Make-believe play was identified as the leading activity of preschool (3–7) age following the adult-mediated object-oriented activity of toddlers and followed by the learning activity of elementary grades students.

Consistent with the major tenets of the Cultural-Historical approach, Vygotskians view play in its cultural and historical context: as children across cultures differ in their social situations of development so does the function of play in their development. In pre-industrial societies, for example, the function of play was mostly to prepare children for engagement in well-defined "grown up" activities. In contrast, play of today's children is non-pragmatic: it does not prepare the child for specific skills or activities, but does prepare the child's mind for the learning tasks of today as well as future tasks that humans cannot yet imagine (Elkonin, 1978, 2005). The Vygotskian view of play through this Cultural-Historical lens means that play cannot be viewed as something that spontaneously emerges as a child reaches a certain age. Instead, play is something that is co-constructed by a child in interactions with other people in the way determined by this child's culture.

The changes in the current culture of childhood seem to be happening across the globe and these changes call into question the status and the future of child play. In this chapter we will not discuss the factors affecting play and how not having "fully developed" or "mature" play affects multiple areas of child development since we have discussed it in detail elsewhere (Bodrova & Leong, 2011, 2018; Leong & Bodrova, 2012). In the context of Cultural-

Historical theory, however, we will mention two aspects that we consider important. First, it is the changes in the social interactions that are beneficial for the development of play. While multi-age play groups with older children functioning as play mentors used to be a common feature of the culture of childhood in many Western countries, today these kinds of interactions become less and less common, which results in fewer children reaching the level of "fully developed" play by the end of their kindergarten year (Miller & Almon, 2009; Russ & Dillon, 2011; Smirnova & Gudareva, 2004). Designing learning environments to allow for multi-age interactions and teaching early childhood educators how to scaffold make-believe play might help children reach the level of play where it truly becomes their leading activity.

The other challenge to the development of play is the changing nature of the mental tools children acquire when engaged in this activity. As we wrote earlier, Vygotsky viewed toys and play props as the tools helping children develop imagination and abstract thought. The use of these toys and props described by Vygotsky and Elkonin followed the path from more realistic to more imaginative, culminating in an abbreviated use of pretend actions as children spent more and more time discussing and planning their play scenarios. Toys used by today's children at home and in an Early Childhood classroom do not allow for their imaginative use, being mostly miniature copies of real objects, often complete with movements and sounds. Supporting's children use of multi-functional and unstructured materials in their make-believe play might in our opinion prevent many of the difficulties these children experience as they start formal education.

Vygotskian Views of Teaching and Learning

Responding to the developmental theorists of his time who claimed that children can learn only those skills and concepts for which they are ready, Vygotsky suggested that this readiness itself can be determined and promoted through the processes of teaching and learning. Arguing with the proponents of "following the child's lead," he writes:

> The old point of view ... assumed that it was necessary to adapt rearing to development (in the sense of time, rate, form of thinking and perception proper to the child, etc.). It did not pose the question dynamically. The new point of view ... takes the child in the dynamics of his development and growth and asks where must the teaching bring the child.
>
> *(Vygotsky, 1997, p. 224)*

Vygotsky agreed that some learning cannot occur until the developmental prerequisites are in place, such as in the case of children not being able to write until their motor skills allow them to have sufficient control of a writing instrument. However, the opposite is also true: certain developments in cognitive, social, or language areas cannot simply emerge as a result of maturation but rather depend on what a child learns. Vygotsky's views of the relationship between learning, teaching, and development remain amazingly relevant today as Early Childhood educators often feel torn between following developmental theories they learned in school and having to implement increasingly demanding educational mandates.

From Vygotsky's perspective, this new point of view calls for a different approach to education, an approach that focuses teaching not on the competencies and skills already existing in a child, but on the competencies and skills that are still "under construction" – the ones that exist in the child's Zone of Proximal Development (ZPD).

Vygotsky's idea of the Zone of Proximal Development or ZPD reflects the complexity of the relationship between learning, teaching, and development as well as the dynamics of the transitions from shared forms of higher mental functions to their individual forms:

> what we call the Zone of Proximal Development … is a distance between the actual developmental level determined by individual problem solving and the level of development as determined through problem solving under guidance or in collaboration with more capable peers.
>
> *(Vygotsky, 1978, p. 86)*

The concept of ZPD is probably the best-known part of Vygotsky's scientific legacy and over the last few decades many scholars have attempted to integrate it into their own theories, offering various interpretations of the idea. In this chapter, however, we will limit our discussion of the ZPD to its original definition and its place in the Cultural-Historical frame of thought.

Vygotsky conceived children's development at any given time as a continuum of skills and competencies at different levels of mastery – hence the word *zone*. By using the word *proximal*, he pointed out that this zone is limited to those skills and competencies that will develop in the near future or are "on the edge of emergence" and not *all* possible skills and competencies that will eventually emerge. The lower boundary of a child's ZPD is defined by the child's level of independent performance and its upper boundary represents his or her assisted performance.

The closer a skill or a competence is to the level of independent performance, the less assistance or support is needed to bring the skill to the surface. When even with the highest amount of assistance a child fails to master a skill or a concept, it means that this particular skill or concept currently lies outside the child's ZPD. The skills and competencies contained within ZPD do not determine children's current level of development but rather their developmental potential. In the absence of guidance or collaboration with more competent others, this potential might not be realized and consequently a higher developmental level will be never attained.

Discussing using the child's ZPD to teach, Vygotsky argued that the most effective teaching is aimed at the higher level of a child's ZPD, which means that teachers should provide activities just beyond what children can do on their own but within what they can do with assistance. While this idea soon became popular among educators, its practical implementation in the classroom met with some challenges: children in a single classroom can differ in their levels of independent performance and the amount of assistance they need, which makes the teacher's task of individualizing quite daunting. In addition, children's individual ZPDs can also differ across different areas of development or for learning of different tasks.

These challenges can be overcome if one takes a broader view of the ZPD, placing it in the context of Cultural-Historical theory. From this perspective, it is not necessary to limit teaching and learning within ZPD to one-on-one teacher–child interactions. It is true that these interactions are necessary to periodically assess the child's ZPD and to help the child acquire mental tools needed to master a specific task or to develop a certain skill. Subsequent mastery of this tool by the child would no longer require frequent teacher assistance but would nonetheless ensure the child learning within the ZPD.

An example would be teaching a kindergartner how to use an alphabet chart during writing. Here the alphabet chart serves as a temporary tool, facilitating the child's mastery of sound-to-symbol correspondence. While trying to remember letters without this tool would result in rudimentary writing (and often in the child's frustration), writing at a higher level when using the tool does not call for constant teacher assistance. Similar to many other mental tools, the alphabet chart would be abandoned once children learn all of the sound–symbol correspondences. Thus, targeting multiple children's ZPD is not a totally unrealistic proposition, but to accomplish this a teacher needs to identify a correct tool that children can use independently. In a sense, when a correct tool is used, this tool acquires part of the functions initially carried out by the teacher and by using the tool the child self-scaffolds, i.e., scaffolds emerging skills and concepts without direct assistance provided by the teacher. Note how this – very practical – example exemplifies the Vygotskian view of the development of higher mental functions from inter- to intramental.

Elementary teachers already use many of these tools (although they may not think of them as tools): times tables, dictionaries, etc. Finding tools for younger children is harder but not impossible. One of the first tools children as young as 2 years start using to regulate their behaviors is private speech, which is audible, self-directed speech (Vygotsky, 1986). Encouraging children to use private speech will help them take control of their behaviors – first physical and later socio-emotional and cognitive.

Another implication of looking at the ZPD from the Cultural-Historical perspective is viewing mature make-believe play as the context where young children get a chance to function at the higher level of their respective ZPDs:

> In play the child is always behaving beyond his age, above his usual everyday behavior; in play he is, as it were, a head above himself. Play contains in a concentrated form, as in the focus of a magnifying glass, all developmental tendencies; it is as if the child tries to jump above his usual level. The relationship of play to development should be compared to the relationship between instruction and development ... Play is a source of development and creates the zone of proximal development.
>
> *(Vygotsky, 1978, p. 74)*

While the description of other sources and mechanisms involved in creating children's ZPDs remained relatively vague in his other writings, Vygotsky's statement on play as a source of ZPD was specific enough to lead to a series of experimental studies, confirming that young children's performance in play was indeed "a head above" their performance outside of play.

For example, Manuilenko (1975) and Istomina (1975) found that young children can perform certain cognitive as well as non-cognitive tasks at a higher level during play than during other activities, which means that they operate at what Vygotsky identified as the higher level of the ZPD. Manuilenko, in particular, found higher levels of self-regulation of children's physical behaviors in play contexts than in non-play contexts. For example, when a child was asked to pretend to be "the lookout," the child remained at his post and did not move for a longer period of time than when the experimenter asked him to stand still in a laboratory condition. In another study, Istomina (1975) compared the number of words children could deliberately remember to put on a shopping list during a dramatic play session involving a grocery store with the number of words they could remember in a typical laboratory experiment.

These findings support Vygotsky's view of play as the "focus of a magnifying glass," as in the quotation above, indicating that new developmental accomplishments do become apparent in play well before they do in other activities. Vygotsky maintained that for children of preschool and kindergarten age their mastery of academic skills is not as good a predictor of their later scholastic abilities as the quality of their play. In a 4-year-old's play one can observe higher levels of such abilities as attention, symbolic thinking, and problem solving than in other situations – one can actually watch the child of tomorrow.

Thus, engaging the entire Early Childhood classroom in a mature make-believe play would target *all* children's highest levels of ZPD. Supporting make-believe play in young children means making sure that children are able to create an imaginary situation and develop a play scenario, can use props in a pretend way, know how to "stay in a character" they chose, and know how to coordinate all these with the other players. Since it is likely that today's children would not learn these skills at home or on a playground, Early Childhood educators will need to take on a more active role in scaffolding make-believe play (Bodrova & Leong, 2011; Elkonin, 1978; Leong & Bodrova, 2012)

Early Childhood Education: What Would Vygotsky Say?

Based on the works of Vygotsky and his students, we can extract some principles of the Cultural-Historical education philosophy as it applies to Early Childhood education.

Early childhood is a cultural phenomenon and its duration keeps getting extended over the history of humankind. It is associated with increasingly more complex tasks facing adult members of society; these tasks requiring the development of complex mental processes not yet directed at solving practical problems. Therefore, early childhood has its own unique value and cannot be viewed as merely "preparing" children for school or for adult life.

Early Childhood educators focus on promoting the development of higher mental functions and on children's acquisition of mental tools (linguistic and non-linguistic) and not on children's mastery of discrete skills and concepts. Specific content knowledge and skills (e.g., writing or counting) are taught to promote these developments and are not considered the end in itself. Although not specifying the exact pedagogy of early instruction in pre-academic skills, Vygotsky outlined its general path, emphasizing that "teaching must be set up so that reading and writing satisfy the child's need" and that the goal of the instruction should be "to teach a child written language and not writing the alphabet" (Vygotsky, 1997, p. 147). Focus on letter formation and learning of the alphabet leads, in Vygotsky's view, to children merely mastering the "writing habit" as opposed to using written language the way it is supposed to be used in culture: as a means of communication with others and as a mental tool supporting one's memory and thinking.

Early Childhood educators promote and foster development by engaging children in activities that are the leading ones for their age (such as emotional interactions for infants, adult-mediated, object-oriented play for toddlers, and make-believe play for preschoolers and kindergartners). This focus on the developmental accomplishments leads to a new definition of school readiness: it is now viewed in terms of supporting the emergence of developmental accomplishments that will ensure children's ability to make the transition to the leading activity of the elementary grades – learning activity. The most important developmental accomplishments of early childhood include self-regulation, imagination, and ability to operate with symbols.

Early Childhood educators constantly revise and adjust their practices to make sure that they target the Zone of Proximal Development of their students. They assess individual and group ZPDs using dynamic assessment. Dynamic assessment is an assessment methodology based on the Vygotskian idea of ZPD. It involves incorporating adult assistance into the very assessment procedure in the form of hints, prompts, or rephrasing the test questions. This modification allows for assessing not only children's existing skills and competencies but also the ones that have not yet surfaced due to the lack of educational opportunities but nonetheless have a potential to develop (Vygotsky, 1986). Dynamic assessment is currently used in psychological evaluation and special education and also proves more accurate in predicting students' academic achievement than traditional formats of assessment (Haywood & Lidz, 2006). Unfortunately, dynamic assessment is not yet commonly used in Early Childhood classrooms where its application seems especially needed.

Early Childhood educators scaffold children's learning and development by first designing and then following a plan for providing and withdrawing appropriate amounts of assistance at appropriate times. Although not used by Vygotsky himself, the concept of scaffolding (Wood, Bruner, & Ross, 1976) explains how aiming instruction within a child's ZPD can promote the child's learning and development. For most children, the transition from assisted to independent is a gradual process that involves moving from using a great deal of assistance to slowly taking over until eventually no longer needing any assistance. From other-assistance to self-assistance and finally to independence, designing appropriate scaffolding means planning to start withdrawing support from the very moment this support is first provided.

Finally, the entire Vygotskian philosophy of early childhood education can be summarized by the concept of *amplification of development*. This term was coined by Vygotsky's colleague and the founder of the All-Soviet Institute for Preschool Education, Alexander Zaporozhets, as the answer to the push down curricula, which resulted in turning the preschool classroom into a miniature copy of a primary classroom with teaching methods and materials modeled after the ones used by elementary teachers. The idea of amplification was also intended to offer an alternative to the notion of "spontaneous development" of young children, the idea that development could not and should not be affected by instruction. Amplification focuses on the role of education in child development, emphasizing that properly designed educational interactions do not stifle development of young children but instead promote it, thus, presenting a logical extension of Vygotsky's principle of instruction leading development.

> Optimal educational opportunities for a young child to reach his or her potential and to develop in a harmonious fashion are not created by accelerated ultra-early instruction aimed at shortening the childhood period, that would prematurely turn a toddler into a preschooler and a preschooler into a first-grader. What is needed is just the opposite – expansion and enrichment of the content in the activities that are uniquely preschool: from play to painting to interactions with peers and adults.
>
> *(Zaporozhets, 1978, p. 265)*

References

Bodrova E., & Leong, D. (2011). Revisiting Vygotskian perspectives on play and pedagogy. In S. Rogers (Ed.), *Rethinking play and pedagogy in early childhood education: Concepts, contexts, and cultures* (pp. 60–72). London and New York: Routledge.

Bodrova, E., & Leong, D.J. (2018) Tools of the mind: A Vygotskian early childhood curriculum. In M. Fleer & B. van Oers (Eds.), *International handbook of early childhood education* (pp. 1095–1111). Dordrecht: Springer.

Elkonin, D. (1972). Toward the problem of stages in the mental development of the child. *Soviet Psychology*, 10, 225–251.

Elkonin, D.B. (1978). *Psikhologiya igry* [Psychology of play]. Moscow: Pedagogika.

Elkonin, D.B. (2005). Chapter 1: The subject of our research: The developed form of play. *Journal of Russian and East European Psychology*, 43(1), 22–48.

Gal'perin, P.I., & Georgiev, L.S. (1960). Psikhologicheskij analiz sovremennoj metodiki obuchenija nachalnim matematicheskim ponjatijam [Psychological analysis of modern methods of teaching concepts in elementary mathematics]. *Doklady APN RSFSR*, 1, Moscow.

Haywood, H.C., & Lidz, C.S. (2006). *Dynamic assessment in practice: Clinical and educational applications.* Cambridge: Cambridge University Press.
Istomina, Z.M. (1975). The development of voluntary memory in preschool-age children. *Soviet Psychology*, 13(4), 5–64.
Karpov, Yu. V. (2005). *The neo-Vygotskian approach to child development.* New York, NY: Cambridge University Press.
Kravtsova, E.E. (2006). The concept of age-specific new psychological formations in contemporary developmental psychology. *Journal of Russian & East European Psychology*, 44(6), 6–18.
Leong, D.J., & Bodrova, E. (2012) Assessing and scaffolding make-believe play. *Young Children*, 67(1), 28–34.
Leont'ev, A.N. (1981). *Problems of the development of the mind.* Moscow: Progress Publishers. (Original work published 1959.)
Lisina, M. (1985). *Child–adults peers: Patterns of communication.* Moscow: Progress Publishers.
Manuilenko, Z.V. (1975). The development of voluntary behavior in preschool-age children. *Soviet Psychology*, 13(4), 65–116.
Miller, E., & Almon, J. (2009). *Crisis in the kindergarten: Why children need play in school.* College Park, MD: Alliance for Childhood.
Russ, S.W., & Dillon, J.A. (2011). Changes in children's pretend play over two decades. *Creativity Research Journal*, 23, 330–338.
Smirnova, E.O., & Gudareva, O.V. (2004). Igra i proizvol'nost u sovremennykh doshkol'nikov [Play and intentionality in modern preschoolers]. *Voprosy Psychologii*, 1, 91–103.
Venger, L.A. (1988). *The origin and development of cognitive abilities in preschool children. International Journal of Behavioral Development*, 11(2), 147–153.
Vygotsky, L.S. (1967). Play and its role in the mental development of the child. *Soviet Psychology*, 5(3), 6–18. (Original work published 1966.)
Vygotsky, L. (1978). *Mind in society: The development of higher mental processes.* Cambridge, MA: Harvard University Press.
Vygotsky, L.S. (1986). *Thought and language.* Cambridge, MA: MIT.
Vygotsky, L. (1994). The problem of the cultural development of the child. In R.V.D. Veer & J. Valsiner (Eds.), *The Vygotsky reader* (pp. 57–72). Cambridge, MA: Blackwell.
Vygotsky, L. (1997). *The history of the development of higher mental functions* (M.J. Hall, Trans.) (vol. 4). New York, NY: Plenum Press.
Vygotsky, L. (1998). *Child psychology* (vol. 5). New York: Plenum Press.
Wood, D., Bruner, J.C., & Ross, G. (1976). The role of tutoring in problem solving. *Journal of Child Psychology and Psychiatry*, 17, 89–100.
Zaporozhets, A.V. (1978). The significance of early periods of childhood for the formation of children's personality. In L.I. Antsyferova (Ed.), *The principle of development in psychology.* Moscow: Nauka.

7

ANTHROPOLOGICAL CONTRIBUTIONS TO EARLY CHILDHOOD EDUCATION

On Culture, Context, and the Construction of the Child

Sally Campbell Galman

In a world where one can study anything, I choose to study children, those small people from whose experience we might most clearly grasp the shape and depth of human learning – if we might bracket our adultist tendencies toward nostalgia, that is. An anthropology of childhood, which is drawn from literature across sociology and anthropology, teaches us that children are people *now*, "human beings not human becomings" (Qvortrup, 1994), and so much more than steps on a developmental ladder. Most importantly, in an era where the United States alone has failed to ratify UNCRC, where children and adults are segregated and processed in factory-like settings, where the fetus is worshipped but the child left to starve in arguably the richest country in the world, that ethnography compels the researcher to position themselves as a humble learner is powerful, indeed. However, the field of early childhood education as a whole has not always valued children as able participants, or even as reliable informants about their own lives and experiences. Further, in research in early childhood, it is typically developmental psychology that has been the disciplinary go-to rather than anthropology or other social science disciplines for the simple reason that developmental psychology provides a normative yardstick against which a child can be measured, for good or for ill.

However, ignoring the yardstick for a moment, Hirschfeld (2002) observed this pattern and asked, "Why don't anthropologists like children?" As Hirschfeld maintains:

> In the briefest terms, mainstream anthropology has marginalized children because it has marginalized the two things that children do especially well: children are strikingly adept at acquiring adult culture and, less obviously, adept at creating their own cultures.
>
> *(Hirschfeld, 2002; p. 611)*

And I would argue (and Hirschfeld would agree) that anthropologists really *do* like children, but the pattern that generated his question remains largely unremarked upon: "Few major works in anthropology focus specifically on children," he writes, "a curious state of affairs given that virtually all contemporary anthropology is based on the premise that culture is learned, not inherited" (Hirschfeld, 2002, p. 611). Curious indeed. As Hirschfeld reminds us, there has been a lot of inquiry around children – pages and pages of it, in fact – however this has "not coalesced into a sustained tradition of child-focused research" in anthropology

(p. 611). It leads one to wonder if it is not a matter of anthropologists not liking children, or anthropological research in childhood simply not gaining a foothold, but instead that anthropologists have been locked out of a conversation dominated by developmental psychology for so long. However, even considering this pattern, this small body of works in the anthropology of childhood *has* had an impact on early childhood education. Even though descriptive, child-centered scholarship does not offer the measurable ease of the developmental yardstick, it has made significant impact upon early childhood practice in general.

While the field of the anthropology of childhood writ large, and its contributions to early childhood education specifically, are a vast and churning sea exceeding the scope of this short discussion, this chapter will offer a few morsels to chew on: What follows is a discussion of the context of a few of those contributions. Namely, that anthropological thought and inquiry have contributed to refocusing the position of the young child and child-centeredness in relation to adults, and also disrupted the dominance of developmental psychology as the primary frame for understanding how the youngest children learn and grow.

Anthropological Foundations

Cultural anthropology is, put simply, the systematic study of human cultures using ethnographic research methodology. Or, perhaps, this is not so simple after all. As Faubion (2007) writes:

> It must be noted at the outset that "cultural anthropology" is no less crisp or constant a category than "culture" itself … [incorporating] the standard of "participant observation," however paradoxical that standard may be. [It demands] that fieldworkers gain some measure of fluency in the languages which their interlocutors natively speak … that they spend time enough among their interlocutors to acquire a sense of what Malinowski called "the imponderabilia of everyday life" and … that they attend to what their interlocutors say, to what they profess to believe and value, and to what they actually do.
>
> *(pp. 39–40)*

The anthropology of childhood is a comparatively new area of study focused on how children learn to be members of, and participate in, culture. As noted above, the study of how children learn has historically been situated in the discipline of psychology, and has focused less on agentive processes of cultural engagement but rather on the application of "culture-bound development theories" (Levine & New, 2008 p. 5) and the child's supposed internalization of adult knowledge regimes, swallowed whole without so much as a gustatory moment. In examining the anthropology of childhood in this context, we might think about Hirschfeld's question, and our own understanding of childhood, and how these require us to step outside the familiar perspectives of classical developmental psychology and its established patterns. A study of children's culture, in the context of Faubion's (2007) definition of cultural anthropology and ethnographic inquiry, requires that we work from the ground up in a kind of radical child-centeredness: One must become conversant in children's discourses, spend significant time immersed in their cultures, and attend to their beliefs and actions with the dedicated focus of serious inquiry. This is a marked departure from a more top-down application of adult-generated models of human development, especially considering that these led to the efficiency discourses, batch processing regimes, and the widespread Taylorism that defined (and continues to define) so much of Western schooling. This is becoming more, rather than less, true even in early childhood contexts as they are dominated by pushdown politics and accountability regimes instead of play-based settings (Galman, 2017; Shepherd & Smith, 1988).

In contrast, ethnography assumes a level of agnosticism – even humility – in its open-ended and discovery-oriented approach to childhood research. Ethnography may not throw out the developmental yardstick, but it actively troubles its normative power. An ethnographer of childhood more often sets out to learn from the child participant, not simply to confirm existing theories of development.

Ethnography relies primarily on participant observation and other forms of eclectic data gathering to create an holistic, descriptive account of what it is like to belong to the culture that is the object of study. The researcher adopts a wholly emic perspective, foregrounding and privileging the participant experience and point of view, and assuming the role of the humble learner. Ideally, this emphasizes the value and complexity of participant culture and experience. Implicit in this is an interrogation of adultism and power, both of which can be tricky in a research endeavor in which the adult holds total power by near default; Indeed, everything from "child saving" and protection/safety rhetorics to nostalgia are used to justify adult belief in children's developmental (and human) incompleteness (Galman, 2017; James, 2007). And children live and function in the context of adult power as adults who believe – with varying degrees of benignity – that children are not fully human, "make most of the decisions for them":

> Unlike adults, who can choose to avoid situations that they find uncomfortable or threatening, children are constantly challenged to develop competence in settings over which they have very little control. A child who finds she does not like her kindergarten classroom cannot, like her older sisters away at college, "change majors." Furthermore, children are rarely allowed the luxury of refusing participation in research or most other adult-conducted activities; again, adults are the gatekeepers. The nature of contemporary children's lives … is that they are constantly under the watchful eye of adults. Children rarely are given private places to work and play. Teachers and caregivers are told that they must be able to see all the children all the time. The boundaries of children's experiences are patrolled by adults.
>
> *(Graue & Walsh, 1998, p. 12)*

Ethnographic research demands confronting this location of power, and admittedly this can be a tall order in most Western contexts, especially the United States. To make a very long story quite short indeed: much mainstream US culture is adultist in the extreme, as legal scholar Roger Levesque (2014) writes:

> The once dominant conception that children essentially were parental property certainly has faded, but … the related principle that parents control their children, with minimal state intervention, remains alive and well … with the United States remaining an exemplary staunch supporter of this approach … This approach, importantly, does not mean that children do not have rights; it does mean that the extent to which they do … largely depends on their parents' decisions to respect them.
>
> *(p. 40)*

American culture is thoroughly saturated with the spirit of what Levesque refers to as *patria potestas*, or the total power of the adult over the child. Paradoxically, the United States is also a modern neontocracy "at odds with virtually every pre-modern society where childcare has been studied" (Lancy, 2015, p. 29). However, when considering the fact that many American children are projects rather than people, this paradox becomes less stark; American neontocracy may actually be more of a display of *patria potestas* than a legitimate culture of child worship.[1] It is against this cultural backdrop that early childhood educators have grappled with the question of autonomy, personhood, and functionalism, and this may be one of the reasons the application of a developmental yardstick seems so promising, which make it difficult to frame much of the messiness of power and context. After all, a functionalist approach in education emphasizes future orientation – the idea that early childhood education is an investment in and concrete preparation for future employment and civic participation. It has its roots, at least somewhat, in Tayloristic efficiency and not so much in children as people now. While early childhood education is populated by more than simply children, and educators, policy makers, administrators, families and others all occupy distinct positions with different needs and orientations to the machinations of schooling, the child is arguably the center, the thing being measured, the person being enculturated through schools and schooling.

James (2007) suggests that ethnography is not only a good and appropriate, but also an inherently emancipatory, fit for childhood inquiry, for these precise reasons and in this precise climate. She argues that: "It is the use of ethnography as a research methodology which has enabled children to be recognized as people who can be studied in their own right within the social sciences" (p. 246). She continues,

> In this sense ethnographic methods have permitted children to become seen as research participants and, increasingly therefore, it is ethnography which is fast becoming a new orthodoxy in childhood research … for what ethnography permits is a view of children as competent interpreters of the social world. This involves a shift from seeing children as simply the raw and uninitiated recruits of the social world to seeing them as contributing to it, a changed perspective which has steered researchers toward doing work "with" rather than "on" children.
>
> *(p. 246)*

The central tenets of an anthropology of childhood, then, are rooted in such emancipatory epistemologies, the implications of which are far-reaching for early childhood education in general for the simple reason that they focus on the child as a competent actor instead of a problem to be solved, a wheel to be oiled, a body to be managed (Galman, 2017). It foregrounds an "interpretive understanding [which] evolves but slowly through immersion in the lives of those we seek to understand" as "ethnography expressly facilitates the desire to engage with children's own views and enables their views and ideas to be rendered accessible to adults as well as to other children" (James, 2007, p. 247). Similarly, there is a tendency in social science research, as well as in early childhood education, to attempt to extrapolate children's experiences and meaning via simply talking to the adults in their lives and taking advantage of the deceptive ease of adult communicative competence. However, as James (2007) reminds us,

1 Lancy (2015) writes that as many cultures, including the United States, have shifted gradually from gerontocracy to neontocracy, "the neontocracy has, lately, gotten out of control. For example, it has been 'argued that the rights of the fetus have now come to supersede the rights of pregnant women themselves' (Landsman, 2009, p. 53)" (p. 71).

> it is not sufficient simply to observe adults' behavior toward children; it is important also to see children as social actors in their own right, to observe and understand what it is that children do with one another as well as with their adult care-takers and, most importantly, to canvass children's own views and opinions directly.
>
> *(p. 250)*

The anthropology of childhood applies the tenets of naturalistic inquiry to the study of children and in doing so positions the child as the knower, and the adult as the one who must learn.

> Ethnography, then, has been critical to the development of a perspective on childhood which, in acknowledging its culturally constructed character, enables a view of children as social actors who take an active part in shaping the form that their own childhoods take.
>
> *(James, 2007, p. 249)*

As noted earlier, the cultural anthropology of childhood draws from literature across sociology and anthropology – indeed, in many European and other contexts, a "sociology" of childhood is much more coin of the realm than is anthropology, which is often seen as much more in the "camp" of physical anthropology, human geography, and even archaeology. However, more often than not, cultural anthropology of childhood is more aligned with childhood studies and comparative study of childhood growth, learning, and development than any of these. Any attempt to tease apart the differences between sociology and anthropology of childhood will certainly reveal areas of overlap, commonality, and cross-pollination; however at its most basic – and for my purposes here – the key difference is the focus on culture, as described above, and the use of ethnography as the primary mode of data collection. To this one must add an emphasis on participant observation as the central methodology. Sociologists also employ ethnography as a methodological tool, to be sure, but it is not considered as field-defining as it is in anthropology, and its use in sociology may be a bit varied from its original purpose among ethnologists and anthropologists: as a method for learning culture and practice from living people (Hammersley, 2010). There are of course numerous variations on this definition, as well as numerous attempts at turning it on its head, but at the most basic, cultural anthropologists of childhood trade in culture (and within that, usually, culture and learning, development, and processes of enculturation), while sociologists may be interested in many aspects of society and social organization. It is as complicated as teasing apart the spaces between "enculturation" and "socialization" – with overlapping, dovetailing ideas that are, and should continue to be, in conversation with one another.

Development in Context

While I do not want to perseverate upon the gulf between developmental psychology and anthropological approaches to understanding early childhood, it would not be fair to ignore both the history of tension between the two, and the dominance of the former in early childhood teaching and learning (Kampmann, 2014). This tension can be viewed as primarily between context vs. normative impulses, tensions which are also present in early childhood practice. As Levine and New (2008) write, the cross-cultural and wide-ranging comparative anthropology of childhood "reflects a critical reaction to proposals of universal standards for child rearing based on the presumption of one 'normal' pathway grounded in the human genome." They continue:

> The more professional experts on child rearing propagated universal concepts of the normal child, the clearer it became to anthropologists that the concepts were fashioned from local (that is, Euroamerican or Western) moral standards combined with biological speculation. To counter this ethnocentric perspective, they conducted field work abroad and provided "de-centered", multicultural understanding of childhood and child development.
>
> *(p. 5)*

Early childhood education has historically been defined by developmental psychology and its maps of how people move from infancy to adulthood, learning everything from how to grasp a small object to how to think about abstract concepts. Scholars have always been interested in learning, but equally interested in normative pattern-making. American schooling in particular made use of these patterns as a means to think about making early learning more efficient – which is to say, that despite the fact that most learning in nearly every culture and throughout recorded history took place in multi-age, multi-generational cross-contextual conditions (Lancy, 2015), most if not all Western schools seek to employ decontextualized models of learning science to maximize outputs in the knowledge factory.[2] Conception, then, builds function: the American way of building and operating schools would probably look very different if the understanding of children and learning had not been predicated on staging and sorting, and had "the European-based system of formal 'Western' schooling [been] seen as a key tool for civilizing" the savage and barbarian (Rogoff, 2003, p. 19). Such perseveration on racist models of linear cultural evolution, combined with adultism and the segregation of children from the adult world, contributed to an American early educational model that has historically been about as far from the approaches of traditional societies worldwide, where children learn from more experienced peers and adults in naturalistic practice, as is humanly possible.

To wit: Developmental psychology is one approach among many to understanding childhood, and its patterns are true and useful for most. However, as Levine and New (2008) and others suggest, we must continue to temper its influence and refine its models with comparative, cross-cultural study. In this way, those models might be refined and made more useful. To all of this one must consider that these models work for "most" children but perhaps not all. For example, the theories of scholars like Piaget are non-universal patterns that can apply to *most* children, *most* of the time, but certainly not *all* children *all* of the time, and that not all of these models accurately account for context, either children's context or the researcher's own.[3] And context, like so much in the lives of children, is ultimately about relationships with power. As Graue and Walsh (1998) remind us, this is particularly pronounced in the lives of young children, who "nevertheless are markedly both more context dependent and context vulnerable than older children and adults" (p. 12). The anthropological emphasis on the difficulties and complexity of context, then, is particularly apropos in the early childhood context.

Further, acknowledging the power and importance of context is one thing, but seeing how it drives normative treatment of human development in concrete ways is a bit more challenging. One particularly good example of how context might shape, or drive, or affect

2 https://www.edweek.org/ew/issues/adequate-yearly-progress/index.html

3 As Rogoff (2003) writes, "Assumptions based on one's own life about what is desirable for human development has been very difficult for researchers and theorists to detect because of their similarity in backgrounds (being, until recently, almost exclusively highly schooled men from Europe and North America)" (p. 20).

developmental ages-and-stages modeling comes from the work of Angulo-Barroso et al. (2011), who were studying motor development in infants. They looked at 9-month-old infants across cultures and how they were learning to grasp both large objects as well as smaller ones. Grasping skills, like sitting up and standing, afford infants with all kinds of opportunities to understand the world and also to engage socially, visually, and cognitively with the world around them. So, the grasping skill represents something quite important. Nine months is the age at which the ages-and-stages literature suggests that most babies begin grasping skillfully. However, as Angulo-Barroso et al. (2011) found, culture has a way of making waves. They discovered that some African infants were unusually precocious in their grasping skills because of cultural care practices that emphasized early supported sitting and standing. One group, studied by Super (1976) in Kenya, is a good example of how cultural practices can affect developmental trajectories in this way: This group was so invested in the value of babies being upright at all times that they placed their infants in specially dug holes, padded with blankets, so that even the youngest child could sit up at all times. These babies learned to sit up, and to grasp, long before typical European babies for whom no sitting ritual was practiced. While the Kenyan babies got ahead quickly, the European babies did eventually catch up, however the point is this: were the normative developmental ages-and-stages scales created in Kenya, the European babies would have been viewed as abnormal. Similarly, American babies from cleanliness-focused white middle-class groups who are not put on the ground to play, but rather held or restrained in devices, learn to walk and grasp much later than those who are allowed to be on the floor (Snow, De Blauw, & Van Roosmalen, 2008). Again, the patterns of developmentalism exist for most babies, and are useful measures, however one must temper their application with appreciation for and acknowledgement of role of context in both human development as well as learning and enculturation. These are things that anthropology does very well, often in concert with developmental perspectives (Rogoff, 2003). The European and American babies are not delayed but simply growing up in a different cultural context. As Super and Harkness (1980) and others affirm, culture shapes development in a mutual and recursive way, and within a dynamic system of interaction between contextual factors, the child is at the center of that process.

So, at its most basic, the anthropology of childhood asserts that development is culturally situated and happens in such a way that it is impossible to separate from the cultural context. So, while one does not reject, wholly, the ideas of Piaget and others who sought out and divined larger patterns in how many children develop and grow, anthropology suggests that we take those with a grain of salt and resist the normative impulse that suggests different types of parenting, learning, child care or even development represent a problem. As Levine and New (2008) write, "Cross-cultural variations in childhood conditions and developmental pathways raise a number of questions, for example, about the effects of early experience and about what infants and the children need for their 'normal' or healthy development" (p. 4). Anthropology certainly draws from the established developmental "ages and stages" thinking, but resists the way in which much early childhood literature treats developmental patterns as monolithic and normative, and how much developmentalist discourse glosses over the Eurocentrism inherent in much of the literature. For example, even the very popular attachment theory falls short of the contextual standard. Attachment theory (Bowlby, 1980), which is predicated upon the idea that infants must develop a strong attachment with a primary caregiver in order to experience healthy emotional and social development, initially failed to accommodate and inadvertently pathologized cross-cultural diversities in family structure, as the model hinges on the premise of a dyadic (usually mother–infant) pairing, rather than more elaborate kinship networks. While this does not mean we simply throw this useful, powerful theory about selfhood and health out the window, one must acknowledge its acontextuality in assuming the relative isolation of mothers and infants in nuclear Western families constitutes a cross-cultural blind spot for attachment theory in particular. Such blind spots do require decentering, as Lancy (2015) comments:

> I've had some success at weakening the intellectual monopoly that western, middle class culture holds on ideas about child rearing and child development. A very thorough review of childhood … from the ethnographic archives has allowed me to offer a cross-cultural and distinctly different account of "normal" childhood.[4]

All childhoods are "normal" childhoods and, as Canella (1997) writes, many strictly developmentalist perspectives "justify categorizing children and diverse cultures as backward and needing help from those who are more advanced" (p. 64).

Children as Agents and Childhood as Complex

Another anthropological contribution to early childhood education is the idea of the child as an agentive participant and reliable informant. Anthropology assumes that children do not simply absorb "adult" culture (or, for that matter, that schooling can use theories of learning and development to speed up, exploit, or otherwise make-more-efficient said process) like swallowing something whole; rather, it assumes that children are capable, intentional users and interpreters of their own cultures and worlds. Corsaro (2003) found that children are "highly skilled producers and directors of their own imaginary worlds [and] active consumers and producers of their own symbolic culture" (p. 91). This is what Nussbaum (1995) might call a generous construction of the child, which is to say seeing the child as able, competent, and fully human. This runs counter to how many early childhood education teachers are themselves culturally conditioned, with a well-intentioned belief in children's incompleteness and vulnerability, and the need to control and curate their worlds for them, either out of a need to protect, to idealize or romanticize, or out of other impulses. However, these instincts are misplaced "and may be located in adults' own truncated view of children's personhood" (Galman, 2017, p. 6). As I have written elsewhere, it is possible that looking to a more generously constructed view of the child could help early years teachers to give children a much wider berth when it comes to play and learning, and, by extension, autonomy and selfhood (Galman, 2017).

However, in addition to being part of the active construction of culture, and the curation of meaning in any given context, the child themselves is a social and cultural construction that is culturally bound. Just as there are multiple childhoods, and childhood itself is not a universal construct, so also the idea of the child and what the child is are contextually dependent. This extends to interpretations of whether or not the category "child" is relevant; When pondering the question of the nature of children and childhoods, Philippe Aries' (1962) notably historical overview of children in pre-modern European society claimed that childhood did not exist historically, and is instead an invention of modern time, writing "in medieval society, the idea of childhood did not exist" (p. 128). His analyses – based largely in art history – suggest that prior to the modern age, adults saw children as disposable, their lives as small adults nasty, brutish, and short, and that adults dealt with increased levels of infant and child mortality by not having much of an emotional connection with them, making little distinction for youth in a marked departure from contemporary (albeit selective)

4 https://hraf.yale.edu/a-cross-cultural-perspective-on-childhood/

Western neontocracy. However, we know that childhood has existed in many forms, and that there is evidence across the historical record that children were afforded special status and given specialized care in many cultures (Orme, 2001). From the discovery of ancient toys to small spoons, furniture, and other objects, and even human remains suggesting that disabled children in some prehistoric contexts were cared for and sustained,[5] we know that the picture of childhoods is complex; similarly, while Aries was chastened for basing his entire claims on the fact that children wore adult-like clothing to have their portraits painted, the idea that childhood is a negotiable and culturally dependent construct remains, and it is expected that this will continue to change over time (James, 2007).

As discussed above, anthropological inquiry into early childhood education recognizes the importance of context as a driving force in development. Learning in context – such as is seen in most traditional societies, and championed uniquely by Vygotsky (1987) – suggests that development is a cultural activity. Vygotsky, while not an anthropologist, was unique among developmentalist thinkers in championing the role of culture and human relationship in learning and development. Like anthropologists, Vygotsky saw the need to interrupt the acontextual, universal visions of the child developing alone and to offer, instead, a descriptive rather than normative vision of learning in a context in which all children are seen as agentive participants in their cultures and communities. Rogoff's (2003) work as a developmentalist building a contextualized model of development puts forward a theory of guided participation to show how children learn by interacting with more knowledgeable others in the dynamic cultural context rather than a static developmental ladder. She writes, "human development is a cultural process … people develop as participants in cultural communities. Their development can be understood only in light of the cultural practices and circumstances of their communities – which also change" (p. 4). She continues,

> To date, the study of human development has been based largely on research and theory coming from middle-class communities in Europe and North America. Such research and theory often have been assumed to generalize to all people … For example, a great deal of research has attempted to determine at what age one should expect "the child" to be capable of certain skills, For the most part, the claims have been generic regarding the age at which children enter a stage or should be capable of a certain skill. A cultural approach notes that different cultural communities may expect children to engage in activities at vastly different times in childhood, and may regard "timetables" of development in other communities as surprising or even dangerous.
>
> *(p. 4)*

Attention to context must trouble these timetables. Historically, anthropologist Franz Boas (1974) used cultural context to decenter the "scientific" rigidity and acceptability of eugenicist claims about race and development by arguing that cultural elements can influence all areas of development, including but not limited to cognitive development, and that because of human plasticity and neoteny we are uniquely susceptible to cultural, social, and environmental influences, resulting in wild variations in normal human development. Like Boas, Mead (1933) and Malinowski (1927) also found that divergence from established Western models of child development did not result in pathology, but instead illustrated that supposed universal models were culturally and contextually informed and dependent.

5 https://www.nationalgeographic.com/science/phenomena/2009/03/30/deformed-skull-of-prehistoric-child-suggests-that-early-humans-cared-for-disabled-children/

Many anthropologists – and others – see Vygotskian ideas about learning in context as a bridge across what Levine (2008) calls the "tricky relationship" between developmental psychology and the anthropology of childhood. Further, a Vygotskian perspective builds upon anthropological perspectives on the child-as-agent, which is to say that just as the child constructs her own understandings, she also constructs her own childhood and participates agentively in cultural production. Too often, researchers and other adults see themselves as necessary curators of children's culture, forgetting the fact that children wield culture with expert precision. To think otherwise is the height of adultism. As Mitchell and Reid-Walsh (2002) write,

> Popular culture, especially mass-media culture, is often constructed as a monolithic giant, while the child is depicted as a powerless object who is about to be consumed. The researchers see themselves as off-screen saviors, rushing in to save the child who is unable to save himself or herself. The researchers, battling and conquering evil, play the role of the prince in the fairy tale.
>
> *(p. 2)*

Children's culture may include the things children do to keep adults at bay, as Graue and Walsh (1998) and James (1989) remind us, and to demarcate the border of the "disorderly and inverted world of children" (James, 1989, p. 404) and that of adults, but it can also be an elaborate place of ritual and meaning. Interestingly, for individuals who spend the vast majority – if not all – of their time under some form of adult control, children often have elaborate and distinct cultures and aesthetics in place. "Children's culture and childhood [are] an independent place with its own folklore, rituals, rules and normative constraints, within a system that is unfamiliar to [adults] and therefore to be revealed through research" (James, Jenks, & Prout, 1998, p. 29). James et al. go on to emphasize that children's culture as the subject of anthropological inquiry is deserving of "detailed annotation" rather than simply the generation of "whimsical tales" (p. 30). To do anything else is to fail to render a meaningful account of children's lives and culture.

Decentering the Adult

Another important contribution anthropology makes to the larger study of children and childhoods is problematizing and recognizing how the field of early childhood education is replete with adultism. As Hirschfeld (2002) writes, most inquiry in childhood has "an impoverished view of cultural learning that overestimates the role adults play and underestimates the contribution that children make" (p. 611). At the core of adultism in early childhood, and even in the anthropology of childhood, is the belief that children are passive recipients of culture. A related, but nonetheless still essential, element of adultism, is the romanticizing of childhoods. For example, anthropology has long asserted that there isn't one single idyllic childhood, but rather a variety of childhoods experienced by children living in different contexts. Moving past the idea that there is a single, normative childhood against which other childhoods are measured (and inevitably come up short) is a major project of anthropologists working to reframe the Western standard of the white, precious cherub and the black and brown children who occupy a marginalized "at risk" or otherwise adultified frame (Ferguson, 2001). Anthropologists assert that all children's experiences are childhoods and all are shaped by the trappings of culture, including but not limited to the cultural and material conditions in those contexts. There is no one normative childhood, and it certainly is not white, wealthy, and Western.[6] And to all of this we must emphasize, again, that children are not simply stewing in the soup of an adult construction of childhood. They are actively constructing their own childhoods. As Frønes (1993) writes: "There is not one childhood, but many, formed at the intersection of different cultural systems, natural and man-made physical environments … different positions in society produce different experiences" (p. 1).

One of many examples of how adultism functions as "blinders" on research in early childhood is in the anthropology of infancy. While some have contested the idea of an anthropology of early childhood, or of infancy, this has been demonstrated to be not only rich but immanently possible, as in the case of Gottlieb's (2004) *The Afterlife Is Where We Come From*. This speaks loudly about the general monolithic adultist, Western belief in the ways in which children speak and how infants cannot be valuable informants about their own lives, but also about the ways in which we might transform our narrow definitions of what counts as communicative competency by decentering our adultism: just because an adult cannot hear or immediately understand does not mean that a child is not communicating. In the case of Gottlieb's Beng infants, they were seen as prolific communicators by the adults and other children in their African community; crying was a language, as was wetting or defecating or failing to nurse (p. 103) and a frustrated parent who could not understand what her baby was saying could only blame her forgetfulness, having lost the ability to understand:

> For many middle-class Western parents … young infants simply lack linguistic abilities. They start out in a prelinguistic phase and only slowly develop the ability to engage intelligently with language … in the Beng view … rather than being either nonlingual or naturally inclined toward language, infants are believed to be multilingual.
>
> *(p. 99)*

The parent is the one who has the problem, for the Beng infant has special knowledge and the adult must work harder to understand them. This rather obviously takes child-centeredness to a bit of an extreme, but this, like so much comparative anthropological work in childhood, gives us stark cross-cultural examples that we might use to problematize and rethink our own ways of thinking about and working with children. How might the early childhood classroom, for example, be different if the teachers and caregivers were to assume not that the children were learning from them, but rather that the children had valuable knowledge, even superior knowledge, and that they and their activities – the imponderabilia – deserved special attention and dedicated study, not simply conformity and compliance?

To build upon this metaphor, and as I have found in my preschool ethnographic work, simply because young children do not have the communicative competencies adults may have developed, they are nonetheless powerful communicators, if only adults would do the work of listening. As linguistic anthropologist Dell Hymes (1974) found, and early educators would agree, there is more to said communicative competence than mere words.

6 For added, and amusing, decentering of the problem of the cherubic, Western childhood, please see Lancy's (2015) "W.E.I.R.D. Societies" – those who are Western, Educated, Industrialized, Rich, and Democratic and whose approach to parenting is at odds with nearly every approach and structure of every non-WEIRD culture or society on the planet.

Conclusion

As James (2007) writes, anthropological inquiry in childhood has been transformative to how childhood is envisioned and treated,

> through close attention to the everyday and familiar through which the social world is both created and sustained, it has enabled the voices of those who would otherwise be silent to be heard … a paradigm for childhood research in which children themselves are regarded as key social actors, whose own views and perspectives are to be taken into account.
>
> *(p. 255)*

With all this emphasis on culture, context, and culturally responsive child-centeredness, the field of early childhood education has begun to bracket its dependence on developmental psychology, and to focus instead on child-centered, contextually responsive ways of working with children and families, due largely to the influence of anthropological approaches in childhood. Anthropology, with its attention to studying the child in the present and in their own right, and in the value of said "imponderabilia," might be called a pioneer in child-centeredness and an important departure from the dominance of developmental psychology, constructing the child as a complete and agentive human being rather than as an incompetent or incomplete step on a ladder to adulthood. The pluralistic, less normative, and more culturally responsive early childhood environment we see in many contexts today could be seen as one product of a cultural anthropological treatment of the child.

References

Angulo-Barroso, R.M., Schapiro, L., Liang, W., Rodrigues, O., Shafir, T., Kaciroti, N., Jacobson, S. W., & Lozoff, B. (2011). Motor development in 9-month-old infants in relation to cultural differences and iron status. *Developmental Psychobiology*, 53(2): 196–210.

Aries, P. (1962). *Centuries of childhood*. New York: Vintage.

Boas, F. (1974). Human faculty as determined by race. In Stocking, G.W. (Ed.), *The shaping of American anthropology, 1883–1911: A Franz Boas reader*, pp. 221–242. New York: Basic Books. (Originally published 1894.)

Bowlby, J. (1980). *Attachment and loss*. New York: Basic Books.

Canella, G. (1997). *Deconstructing Early Childhood Education: Social justice and revolution*. New York: Peter Lang.

Corsaro W.A. (2003) *We're friends, right? Inside kids' culture*. Washington, DC: Joseph Henry Press.

Faubion, J.D. (2007). Currents of cultural fieldwork. In Atkinson, P., Coffey, A., Delamont, S., Lofland, J., and Lofland, L. (Eds.), *Handbook of ethnography* (pp. 39–56). London: SAGE.

Ferguson, A.A. (2001). *Bad boys: Public schools and the making of Black masculinity*. Ann Arbor, MI: University of Michigan Press.

Frønes, I. (1993). Changing childhood. *Childhood*, 1(1), 1–2.

Galman, S.C. (2017). Brave is a dress: Understanding "good" adults and "bad" children through adult horror and children's play. *Childhood*, 24(4), 531–544.

Gottlieb, A. (2004). *The afterlife is where we come from*. Chicago: University of Chicago Press.

Graue, M.E., & Walsh, D.J. (1998). *Studying children in context: Theories, methods and ethics*. Thousand Oaks, CA: SAGE.

Hammersley, M. (2010). Ethnography. In Baker, E., Peterson, P., and McGaw, B., (Eds.), *International encyclopedia of education*, 3rd edition. Oxford: Elsevier.

Hirschfeld, L.A. (2002). Why don't anthropologists like children? *American Anthropologist*, 104(2), 611–627.

Hymes, D.H. (1974). *Foundations in sociolinguistics: An ethnographic approach*. Philadelphia: University of Pennsylvania Press.

James, A. (1989). Confections, concoctions, and conceptions. *Journal of the Anthropological Society of Oxford*, 10(2), 83–95.

James, A. (2007). Ethnography in the study of children and childhood. In Atkinson, P., Coffey, A., Delamont, S., Lofland, J., and Lofland, L. (Eds.), *Handbook of ethnography* (pp. 245–258). London: SAGE.

James, A., Jenks, C., & Prout, A. (1998). *Theorizing childhood*. Cambridge: Polity Press.

Kampmann, J. (2014). Young children as learners. In Melton, G.B., Ben-Arieh, A., Cashmore, J., Goodman, G.S., and Worley, N.K. (Eds.), *The SAGE handbook of child research* (pp. 136–152). London: SAGE.

Lancy, D.F. (2015). *The anthropology of childhood: Cherubs, chattel, changelings*. Cambridge: Cambridge University Press.

Levesque, R.J.R. (2014). Childhood as legal status. In Melton, G.B., Ben-Arieh, A., Cashmore, J., Goodman, G.S., and Worley, N.K. (Eds.), *The SAGE handbook of child research* (pp. 38–53). London: SAGE.

Levine, R.A., & New, R.S. (2008). *Anthropology and child development: A cross-cultural reader*. London: Blackwell.

Malinowski, B. (1927). *Sex and repression in savage society*. New York: Harcourt Brace.

Mead, M. (1933). *Coming of age in Samoa*. New York: Blue Ribbon Books.

Mitchell, C., & Reid-Walsh, J. (2002). *Researching children's popular culture: The cultural spaces of childhood*. New York: Routledge.

Nussbaum, M.C. (1995). *Poetic justice: The literary imagination and public life*. Boston: Beacon Press.

Orme, N. (2001). *Medieval children*. New Haven: Yale University Press.

Qvortrup, J. (1994). *Childhood matters*. Aldershot, UK: Avebury.

Rogoff, B. (2003). *The cultural nature of human development*. Oxford: Oxford University Press.

Shepherd, L.A., & Smith, M.L. (1988). Escalating academic demand in kindergarten: Counter-productive policies. *The Elementary School Journal*, 89(2), 135–145.

Snow, C., de Blauw, A., & van Roosmalen, G. (2008). Talking and playing with babies: Ideologies of child-rearing. In Levine, R.A., and New, R.S. (Eds.), *Anthropology and child development: A cross-cultural reader* (pp. 115–126). Malden, MA: Wiley-Blackwell.

Super, C.M. (1976). Environmental effects on motor development: The case of African infant precocity. *Developmental Medicine and Child Neurology*, 18, 561–567.

Super, C., & Harkness, S. (1980). *Anthropological perspectives on child development*. San Francisco: Jossey-Bass.

Vygotsky, L.S. (1987). Thinking and speech. In Rieber, R.W., and Carton, A.S. (Eds.), *The collected works of L.S. Vygotsky*. New York: Plenum Books. (Originally published 1934.)

8

A SOCIOLOGY OF EARLY CHILDHOOD EDUCATION

Learning to be Civilized

Norman Gabriel

Introduction

This chapter will argue that we need to develop a relational, sociological approach that explains the early learning and educational experiences of young children. I will begin by reviewing some of the contemporary trends in Early Childhood Education and Care (ECEC), before focusing on some of the major sociological approaches to childhood that have been used to understand the socialization and education of young children. As more young children enter childcare and early education, the issue of transition from their immediate families to group care settings has grown in importance. Young children are born into interdependent relationships that existed before them: as they grow up, these relationships with their parents, teachers, and friends change but are structured by different societies in different historical epochs.

I will turn to Norbert Elias and Pierre Bourdieu, two of the most eminent relational sociologists, to explain the changing institutional arrangements of care and education that young children experience. These institutional arrangements set the tone for relationships between individual children, dyads (child–teacher), peer–peer (friendships and playmates), and group interactions between young children, teachers, and caregivers. My argument will be that Elias's concept of "love and learning relationships" provides enormous theoretical potential, enabling us to focus on the cognitive and social relationships of learners and to develop the relational turn that has emerged in other disciplines apart from sociology, particularly in psychoanalysis.

I will then build on the findings from the British object school of psychoanalysis to uncover and explain the emotional anxieties of young children as they learn to become more civilized and internalize from adults and their peers an enormous social fund of knowledge about the world. Bourdieu's theoretical framework will be used to explore the development of a social habitus in early years education, one where young children accumulate their own stocks of social capital through their strategic use of learning networks.

Early Childhood Education and Care

Early childhood is a complex field with many varied terms, including early years, early childhood development, early care, early care and education, and early childhood education and care. While there are no clear definitions, the terms "early childhood" and "early years" are among the two most popular internationally and are often used interchangeably by researchers. Therefore, the two terms "early childhood" and "early years" will be used in this chapter. There is also some controversy regarding the age span of children that should be included in early childhood, with most using the term to refer to young children from birth to age eight. Despite the contested nature of the ages to be included in early childhood, the vast majority of international researchers consider early childhood or the early years as embracing birth to age eight years (Farrell et al., 2015).

Despite this complexity, it is nevertheless possible to identify some common trends and tensions in the different knowledge traditions in early years education. For example, an OECD (2006) investigation of early childhood education in the Nordic countries, in most European countries, and in the United States, Canada, Australia, Mexico, and Korea identified two different approaches to early childhood education. A social pedagogy approach also sometimes referred to as the Nordic and German model is local, child-centered and holistic, and uses concepts including care, play, relationships, activity, and development. This approach views children as agents of their own learning. In contrast, in the UK, Australia, Canada, France, Ireland, and the Netherlands there has developed an early years education approach that is more academic and centralized, strategically oriented toward teaching, learning, curriculum, content, and methodology.

In addition to these different knowledge traditions, there has been an increasing emphasis on early intervention and regulating young children's upbringing and behavior (for example through parenting orders and fining or jailing parents for child truancy) (Lister, 2006). Statham and Smith (2010: 17) have identified three different but related approaches to early years interventions: those which target the pre-natal period or young children of any preschool age; early identification of problems and additional needs; and earlier delivery of services and interventions aimed at promoting resilience among groups at risk of poor outcomes. Programs for early interventions vary in their primary aims, focusing on, for example, education, health, early literacy, childcare, family support, and parenting.

Recently, Sahlberg (2014) has popularized an important debate about the worldwide trend described as "Global Educational Reform Movement" (GERM). Although GERM is not a formal policy program, there are some common features which have been adopted in predominantly Anglo-Saxon countries, such as the United States, England, Australia, and New Zealand. Through international funding organizations, such as the World Bank, elements of this reform have also spread to other countries, especially in the less developed parts of the world (Sahlberg, 2016). These elements include, for example, the standardization of teaching for predetermined learning outcomes by using prescribed curricula. This is attempted by detailing the delivery of lessons and evaluating predetermined measures, with very little consideration of local, cultural, or individual differences (see, for example, Robertson, 2015).

We can summarize this contemporary world-wide trend by referring to terms such as "standards," "accountability," and "effectiveness," which are now dominant in education policy discourse and have replaced "autonomy," "trust," and "pedagogy" (Sahlberg, 2016). Professional autonomy is increasingly replaced by the standardization of schooling and education; standards, pre-defined learning outcomes, prescribed curricula, testing, and accountability are more and more shaping early childhood education and care. According to Moss (2014), there is little space for uncertainty, experimentation, surprise, or amazement.

Schooling and the "Socialized" Child

> Thus by studying education historically we shall be enabled not only to understand the present better, but also we shall have the chance of revising the past itself and of bringing to light mistakes which it is important for us to recognise, since it is we who have inherited them.
>
> *(Durkheim, 1977: 17)*

Although Durkheim's *The Evolution of Educational Thought* remains comparatively unknown to sociologists, his approach is important for connecting the present with the past in a process-oriented approach that captured the way in which, on the one hand, everything is in flux and changeable – yet, on the other hand, education retains its structure through its relation to contemporary society. An important way for understanding some of the changes and continuities in the structure of education is to explore the emergence of schools as a key institution in young children's lives. This historical context can help us to understand contemporary developments in early childhood education by making us more aware that schooling and the curriculum are influenced by powerful adults who make decisions about the ways young children are taught and about how they will spend much of their time.

In the nineteenth and early twentieth centuries, the central place of work in young children's lives was changed by child labor legislation and compulsory education. According to Hendrick (1997), the campaign to prevent young children from working in the factories was one of the first steps in the construction of a universal childhood. This removal of children from the workforce started with the Factory Acts of the 1830s and 1840s, continuing in the latter half of the nineteenth century with the introduction of the Elementary Education Act of 1870, which established a requirement to provide universal elementary education. Policy makers promoted a state-funded education service as a way to relieve poverty and to prevent the spread of civil unrest and crime that was perceived by the middle and upper classes to be developing as a result of the large numbers of children roaming the streets in inner cities (Petrie, 2003). Once these children entered school, local authority officials and Victorian governments in the latter part of the nineteenth century became increasingly concerned that young children were unable to "concentrate" and attend to lessons, due to under-nourishment and infectious diseases. Concerns for children's health and fears that the next generation would not provide "fit" soldiers for war eventually led to the introduction of free school meals, medical inspections, and treatments.

Alongside this concern for fitness and obedience was the government expectation that investment in children would lead to the development of a disciplined and skilled workforce – scripture, reading, writing, and arithmetic were instilled through rote learning. Children were required to write in a "fair hand," often by copying poems from the established canon of English literature or from morally uplifting tracts. They had to learn the discipline of parsing sentences in order to learn the syntax of the English language, and had exercises in the comprehension and paraphrasing of high-status texts.

In "The School Class as a Social System," Parsons (1959) argued that the central functions of education were socialization and selection:

> Our main interest is … first, of how the school class functions to internalise in its pupils both the commitments and capacities for successful performance of their future adult roles, and second, of how it functions to allocate these human resources within the role-structure of the adult society.
>
> *(Parsons 1959: 297)*

In the structural-functionalist writings of the 1950s and 1960s socialization became defined as a psychological process whereby the young child learns the "laid-down" patterns of values that will mold him or her to fit into existing society. Parsons (1951) discusses how the school classroom establishes the conditions through which these values are obtained, including the elimination of distinctions of prior social background, construction of standardized tests to demonstrate aptitude, and role modeling of the teacher, who represents the adult world and ultimately determines which social functions within the division of labor students should strive toward. In this traditional view in the sociology of education, socialization explains the transmission of culture from one generation to another through the key institutions of community, education, and the family (see Elkin & Handel, 1972). Learning to conform to social rules, young children gradually acquire knowledge of the roles needed for adult life – they must be shaped and trained in order to become a competent and contributing member of society.

However, a key weakness in the structural functionalist perspective was the major assumption that the child is "mere putty to be worked on by external forces." Researchers working in the new social studies of childhood that emerged in the 1970s and early 1980s began to rethink childhood and challenge this view that young children were mere passive recipients of socialization (Richards, 1974). Sociologists of childhood (see in particular, James & Prout, 1997) emphasized the present tense of childhood, children's active participation in constructing their own lives, and their relationships with parents and friends. They argued that in the early years of human life a different framework is needed to understand the institution of childhood: "children are not formed by natural and social forces but rather ... they inhabit a world of meaning created by themselves and through their interaction with adults" (James et al., 1998: 28).

Prout (2005) has argued that though researchers working within the new social studies of childhood have been productive, there are intellectual limitations in their research program, which is based upon a set of oppositional dichotomies. Their theoretical framework assumes that childhood is a social construction which stands in opposition to older biologically centered ideas of childhood. This was understandable in the formative stage of the approach's development, because in order to establish their distinct contribution, novel intellectual initiatives frequently overstate their case, emphasizing their differences from previous formulations. However Prout (2011) suggests that the sociology of childhood must move away from dichotomies such as seeing the child as either a "being" or a "becoming" and that such a move needs to investigate the relations between such oppositions.

In an innovative paper, Quennerstedt and Quennerstedt (2014) offer a different approach from that of Prout (2011), arguing that the metaphysical structure of children as "beings" or "becomings" should be avoided altogether. They develop a theoretical approach that integrates aspects of the sociology of childhood with John Dewey's educational focus on growth. Dewey (1938) believed that education should be based on growth in terms of the re-organization and re-construction of experience: "the educational process has no end beyond itself": it is a process "of continual reorganizing, reconstructing, transforming" (Dewey, 1916: 50). This concept of growth is intimately related to and dependent on his other concepts, particularly "continuity of experience," "education," and "democracy," where no single concept is taken in isolation without its connection to others. Dewey's concept of growth is therefore deeply relational and embedded in his theory of educative experience.

Relational Pedagogy

Papatheodorou (2009) argues that development of a relational pedagogy can challenge the dominant performative rationality of ECEC as a market with its objectification of the child, overcoming the dichotomy between an outcomes-based and processes-oriented pedagogical praxis. She suggests that relational pedagogy, understood as a complex web of human experiences rather than an individual experience divorced from its cultural and social context, signals a significant change in the conceptualization of early years education, bridging polarized discourses such as child-centered versus adult-centered learning.

According to Papatheodorou (2009), the dialogical aspects underpinning relationships between learners and teachers offer the opportunity to challenge teaching as a technical act, resisting the imposition of a priori beliefs of who the learner is or should be. Teachers can re-examine their attitudes and understanding of students, making "thinking with" possible. A good example of the type of concepts that underpin this relational form of early years education is the Maori word "*ako*," which means both "to learn" and "to teach," reflecting the reciprocal processes implemented in *Te Whariki* curriculum in New Zealand. These concepts entail a web of "betweenness" that emphasizes the relational dynamics of various contributions that each individual can make.

The educational writings of the Scottish philosopher John Macmurray have a special relevance in developing this relational view of education, one that enables us both to ask questions we too often ignore and develop responses that overcome the more superficial aspects of contemporary educational debate. Writing about these pedagogical matters more than 80 years ago, Macmurray insisted that we should educate the emotions, placing relationships and care at the heart of teaching and learning. We are, in Macmurray's view, deeply and irrevocably relational beings whose creative energies are best realized in and through our encounters with others: he insists on education as a relational, caring undertaking:

> A child is born human; ... He can survive only by being cared for. He can do nothing - just nothing - to help himself. He has to learn everything - to see, to move about, to walk, to speak: and while he is learning these basic elements of humanity, his human life consists in his relation to those who care for him - who feel for him, think and plan for him, act for him. This dependence on others is his life - yet to be human he must reach beyond it, not to independence, but to an *interdependence* in which he can give as well as receive.
>
> *(Macmurray, 2012: 666–667, author's emphasis)*

Macmurray distinguished between three types of knowledge. The first kind of knowledge, which he calls "knowing how" is typified by the sciences. The second kind of knowledge, "knowing why," helps us to determine ends and assign priorities and is typically represented by the arts. What is particularly significant for my argument in this chapter is the addition of a third, even more important, kind of knowledge that was central to Macmurray's work and received its fullest expression in his 1955 Gifford Lectures (Macmurray, 1957, 1961). He refers to this kind of knowledge as "knowledge of community" (Macmurray, 1965), which has important implications for how we view the following relationships: the importance of schools as living communities; the centrality of relationships of care in the educational process (he insisted that we should educate the emotions, placing relationships and care at the heart of teaching and learning); the techniques and methods we use to engage each other in learning; and the curriculum itself through immersion in forms of community life that affirm our mutuality as persons.

Love and Learning Relationships

Although Elias did not explicitly address educational practice or the role of education in society, he was deeply interested in the development of the social learning processes of young children and adults. For young children there are "natural human structures which remain dispositions and cannot fully function unless they are stimulated by a person's 'love and learning' relationship with other persons" (Elias, 2009: 147). This important relational concept of love and learning aptly summarizes a great deal of previous psychological research on young children's development, bringing together specialized areas within psychology (particularly the separation between cognitive, social, and developmental psychology).

In *On the Process of Civilization* Elias writes that

> Since people are more or less dependent on each other first by nature and then through social learning, through education, socialization, and socially generated reciprocal needs, they exist, one might venture to say, only as pluralities, only in figurations.
>
> *(Elias, 2012: 525)*

To highlight the strong, affective ties that link people with one another, Elias also used the notion of *valency* to refer to the relational way in which people are directed toward other people: some are already firmly connected with certain people, while others are free and open, and search for people with whom to form bonds. Elias therefore emphasizes the importance of personal interdependencies and emotional bonds that bind society together: human beings are social beings embedded in figurations which are interdependent webs and networks that are always moving, changing, and developing.

The historical development of processes of civilization has had a dual impact on childhood: first, the distance between childhood and adulthood gradually increases as the requirements of societal membership become more demanding, so that childhood requires more time and effort in socialization and education prior to the achievement of adult status through entry to the workforce. Second, adults' investment of time, skill, effort, and emotions in young children also increases, making them both more "precious" and demanding at the same time. An integral aspect of this civilizing process is that young children should eventually grow up through their own self-regulation. Elias mentions a unique human capacity "for controlling and modifying drives and affects in a great variety of ways as part of a learning process" (Elias, 2007: 125). However, this process is largely forgotten as adults, where a high level of civilizing restraint forms part of their social habitus. This restraint appears to grown-ups as "automatic," a part of their "second nature," which is treated as something with which they were born. I will now use the relation between love and learning as a sensitizing concept to focus on the way in which the education of young children is both a cognitive and affective process, one that can best integrate the findings of psychoanalysis to develop a suitable sociology of early years education.

Relational Psychoanalysis

A distinctive school of relational psychoanalysis developed from the early 1980s in the work of Greenberg and Mitchell (1983), who posited a relational model in opposition to classical Freudian drive theory. This "new tradition" draws on three long-standing bodies of thinking in psychoanalysis: the American interpersonal tradition (see, for example, Sullivan, 1953), which emphasized the importance of understanding the network of relationships within which individuals exist; the British object relations tradition (Bion, 1962; Bowlby, 1969; Winnicott, 1965); and the work of American psychoanalytic feminists (Benjamin, 1998; Chodorow, 1999; Dimen, 2003). According to Roseneil and Ketokivi (2016), each of these lineages of theory posed its own challenges to the monadic model of drive theory, with its primary focus on intra-psychic processes, on the quest for rational control by the ego and the developmental goal of separation and autonomy. Their shared orientation conceptualizes the self as relationally constituted, where the matrix of mother or carer–child relations provides the very conditions of possibility of existence for the young, dependent child. Hence, from the beginning, the self is intrinsically social, our sense of autonomy and agency inherently relational.

I will now focus on two of the major thinkers of the British object relations school, Donald Winnicott and Wilfred Bion, and some of their important concepts for developing a sociology of early years education, one that does not assume that pedagogy is predominantly a rational, conscious, and deliberate process. Rational definitions of pedagogy have traditionally been based on the actions of the pedagogue and proceed from the assumption of building incremental knowledge upon the edifice of the learner. Bibby (2011) argues that this explanation of the teacher/learner is too simplistic and does not take into account how young children are part of a relational, less linear process.

Winnicott (1971) argued that relationships precede individuality and are governed by the need to relate. He wanted to show how young children bridge the gap between egocentricism and recognition of an external world, and how they negotiate and renegotiate relations between self and other. Playing the other enriches and expands the boundaries of the self and at the same time sharpens the differences between the two. It is designed to attach the infant to the caregiver and at the same time enable him to keep the right distance from her. In this space, the baby identifies something, an object, which is at the same time "not me" and "not mother." Using that object is important since only by identifying it as non-self can it act as a transitional object and enable the infant to relate his inner reality to what is outside.

Winnicott introduces the important concept of a transitional space that is "outside the individual, but it is not the external world ... Into this play area the child gathers objects or phenomena from external reality and uses these in the services of inner or personal reality" (Winnicott, 1971: 51).

This potential space must avoid being challenged because it belongs to neither an inner nor an outer reality, but must remain in-between. Winnicott does not separate the child from her environment in terms of the discovery of self, objective distancing, naming, or rationalizing but proposes a fluid process of separation involving intuition, experimentation, and play. This space of pedagogy and creativity is one that cannot be defined by the terms "inner" or "outer" and thus highlights the agency of children and their ability to make use of space in conceptualizing identity, place, and difference. Winnicott's (1950, 1971) concept of transitional space and playful pedagogy therefore provides a radical outlook on the processes of young children's learning.

Learning to be Civilized

> The child, in learning to be civilised, naturally also feels frustrations acutely, and is helped in becoming civilised not so much by the teacher's precepts as by the teacher's own ability to bear the frustrations inherent in teaching.
>
> *(Winnicott, 1964: 202–203)*

Bion (1962) contends that the growing toleration of frustration allows "thinking" to develop. For Bion, the process of thinking and learning is rooted in the developing ability of humans to tolerate "uncertainty" and "unknowing." This process is fraught with the difficulty of staying with the experience of uncertainty. Uncertainty is what education feels like, it involves getting to know one's emotional experience from the pain and vulnerability of learning from ambiguous experience. Bion (2004) calls this "knowledge," arguing that the process of coming to know depends on how we manage to tolerate the frustration of uncertainty – of "not knowing." According to Bion, the conceptualization of experience is quite different from our taken-for-granted view that it is acquired and cumulative.

Another key aspect of Bion's argument is his belief that the toleration of pain and frustration is central to learning since all learning requires that we take a risk. To move from not knowing to knowing we must necessarily move through a period that will involve uncertainty, frustration, and anxiety that inevitably carries risk. Failure and success both entail risk. In discussing Bion's insistence on the centrality of the relationship between learning and experience, Deborah Britzman (2003: 28) comments, "having to learn is [not] an experience that can be known in advance. And this radical uncertainty ... is the structuring tension in education."

Bibby (2009) argues that Bion's work offers a powerful way of thinking about knowledge, knowing, and relatedness and has important implications for understanding the processes of learning for young children in early years education. In one of her studies that examined children's learner-identities in primary school mathematics, she demonstrated the difficulties associated with "refused" or "blocked" relationships: "knowledge about" – curriculum content knowledge – was only acceptable when it was intimately bound to the emotional connections to, and work with, teachers and peers. When curriculum content knowledge was taught alongside a denial of this interconnectedness, it was reduced to unsatisfying, disconnected knowledge. Bibby (2009, 2011) argues that when educational practices are based upon anxieties (for example, tests or exams) institutional mechanisms tend to block the links between learning to know and relatedness.

The Social Habitus of Early Years Education

Elias's important relational concept of love and learning can be further utilized to explore the significance of institutional processes in young children's education, focusing on how from one generation to the next young children internalize from adults and their peers an enormous social fund of knowledge about the world. He argues that though there is a great deal of psychological and physiological literature on learning there is very little on the structuring of the habitus through learning (Elias & Dunning 2008, 93). According to Brooker (2007: 14), learning can be seen as a series of relationships between children and their friends and classmates, between children and the adults who care for them in every setting, and between the professional educators and the families and communities who have provided children's earliest experiences.

I will now apply Bourdieu's relational concepts of field and internalized dispositions in *habitus* as an important theoretical framework for developing a sociology of early childhood education. The social habitus refers here to the internalization of wider structures and processes manifested through the routines and taken-for-granted actions of young children: the longer an individual child is located within a particular set of relationships, the more likely she or he is to develop a practical sense of how to behave and act in certain ways.

Wacquant (2014) argues that we need to elaborate on Bourdieu's distinction between primary and secondary habitus, introduced in his work on education and underlying his analysis of the nexus of class and taste in *Distinction* (Bourdieu, 1999). The primary habitus is the set of dispositions one acquires in early childhood, slowly and imperceptibly, through familial immersion, where young children initially derive modes of thinking and types of dispositions from their parents or carers. These different dispositions that have been transferred during childhood are an education that is linked to the parents' social position in the social space. Social spaces are seen as fields of struggle, caught between forces of transformation and preservation and depending on access to power and resources. Therefore, the primary habitus is about "internalizing the external" as the parents' modes of thinking, feeling, and behaving that are linked to their position in the social space are internalized in the child's own habitus. This is what Bourdieu and Passeron (1977) also refer to as *class habitus* and it reflects the different positions people have in society, leading to different lifestyles tastes and interests among social classes (Bourdieu, 1999).

Wacquant (2014) argues that the primary habitus is a springboard for the subsequent acquisition of a secondary habitus which results from one's education at school and university, but also from other life experiences. However, the primary habitus as "embodied history, internalized as second nature and so forgotten as history" (Bourdieu, 1990: 56) never loses its impact and always influences the development of the secondary habitus. It is also fashioned by tacit and diffuse "pedagogical labour with no precedent"; it constitutes our social personality as well as "the basis for the ulterior constitution of any other habitus" (Bourdieu & Passeron, 1977: 42–46). Pedagogical labor is about transformation, who defines what shape it will take, and how it is experienced in practice by young children.

Peer Groups and Capital Accumulation

We can further develop this concept of the primary habitus by exploring how young children are able to convert and use their own capitals within school and early years settings. These settings can be viewed within the concept of a shifting and competitive field where young children accumulate their own stocks of social capital through strategic use of learning networks. Bourdieu conceived of four different types of capital which are often deeply interrelated and partly transposable: economic capital, "which is immediately and directly convertible into money and may be institutionalised in the form of property rights" (Bourdieu 1986, 243); cultural capital, which consists primarily of what is to be perceived to be legitimate knowledge and behavior and may be institutionalized in the form of educational qualifications; social capital, which relates to the prestige and influence gained through relationships and connections with powerful others; and symbolic capital, which represents the status and honor that is associated with the acquisition of one or more forms of capital once they have been perceived and recognized as legitimate by others.

Although families are still important institutions in the accumulation and transmission of economic, cultural, social, and symbolic capital, young children are not mere receptors of their family's capitals, but active generators through their structural positioning in early years settings. Parents, siblings, friends, childcare providers, and teachers all influence young children's social relationships. What is particularly significant is the role that siblings and peer relationships play in the formation of young children's learning and socio-emotional development. Young children seek, in their peers and friendships, the emotional bonds and feelings of security they first established in families. They value similar dimensions of intimacy, support, trust, and mutuality as older children and adults (Dunn, 2004).

Only a few researchers have examined young children's peer groups and capital accumulation in early childhood institutions from a Bourdieusian framework (see Connolly, 2000, 2004; Palludan, 2007). Palludan (2007), for example, examines how kindergarten children are differentiated and segregated through vocal practices and processes. Her argument is that different groups of young children have unequal access to resources for generating respectability and obtaining recognition in a specific, linguistic market. Bourdieu (1995) argues that linguistic practices should always be interpreted as an encounter between linguistic habitus on the one hand, and the structures of the linguistic market, which are forms of recognition with specific sanctions and censorship, on the other. By using the term "market," Bourdieu (1993) emphasizes that it is not only important to speak correctly in the social space; if your linguistic contribution is to carry weight in a specific context, you need to use socially accepted expressions and master the art of speaking in a way that counts.

Palludan (2007) uses Bourdieu's concept of a "linguistic habitus" to distinguish between two different language tones, a teaching tone and an exchange tone, that were practiced in a Danish kindergarten. The teaching tone can be described as the vocal form that, from the adult's perspective, enables them to establish and maintain a language-based community even if the young child has little or no knowledge of Danish. The exchange tone is typically found in situations where the ethnic majority children (the Danes) and the adults converse – both sides ask questions and give answers. One of her important findings was that adults are continually prone to adopting teaching tones with young children whose first language is not Danish, even when they are quite competent in Danish. Yet at the same time they tend to use exchange tones with young children whose first language is Danish. Despite their best attempts to include all the kindergarten children, the adults contributed to processes that reproduced a socio-cultural hierarchy amongst young children.

In another study Devine (2009) considered how migrant children in Irish primary schools were not merely receptors of their family's capitals but active generators, contributing to processes of capital accumulation through their negotiation and positioning between home and school. Older siblings spent time doing homework with younger siblings – especially helping with Irish – and were also actively involved in caring for younger children when parents worked in the afternoons and evenings. Where parents were not fluent in English, or had difficulty accessing their own social networks, it was the young children who acted as mediators – teaching their parents English, acting as translators, or introducing them to Irish parents.

Devine noticed that friendships were important sources of social capital, facilitating access to learning networks that provided relief from the demands of formal learning, as well as support and knowledge when challenges emerge. These friendships gave young children a feeling of belonging and "getting on" in their everyday lives in school. The "durable obligations" (Bourdieu, 1986: 249) built up through these learning networks ensured that they could draw on their friends to help them with school work, especially homework, as well as "defend" them if they were being racially abused. It therefore seems that in young children's peer groups, especially where economic capital is not directly used, the importance of social capital is emphasized. If a young child has a recognized position in a friendship network, it is easier to obtain access to other capitals that others have already gained in related fields.

Conclusions

This chapter began by providing an overview of the current theoretical debates within the sociology of childhood and early years education and care, contextualizing these within some of the long-term changes and continuities about what counts as "educational knowledge." I have argued that we need to move beyond attempts to define the different knowledge traditions of ECEC, recognizing the limitations of standardized, rational approaches that specify pre-defined learning outcomes by using prescribed curricula. Such a narrow focus on the curriculum highlights the difficulty of finding a linear causality between teachers' actions and young children's learning, providing a distraction from, and excuse not to undertake, the important emotional work required to understand the complexity of learning relationships.

To develop my argument for a sociology of early years education I emphasized the great potential of Elias's concept of "love and learning" relationships. I used the relation between love and learning as a sensitizing concept to discuss how young children's education is both a cognitive and an affective process, one that is fraught with a great deal of emotional anxieties that gradually require more self-regulation. I explored the significance of institutional processes in the social habitus of young children's education, focusing on the way that from one generation to the next young children learn from adults and their peers an enormous social fund of knowledge about the world.

I then turned to two of the major thinkers of the British object relations school, Donald Winnicott and Wilfred Bion to question the assumption that pedagogy is predominantly a rational, conscious, and deliberate process. By refocusing on the unconscious processes of risk and uncertainty in the classroom, we can offer a way beyond the dead-end approach of accepting or rejecting prescriptive advice on how to teach better or how to deal with "difficult" pupils. To develop a sociology of early years education, we should look to Dewey's attempt to capture the uncertainty of the educational endeavor itself and the radical uncertainty of thinking: "All thinking involves a risk. Certainty cannot be guaranteed in advance. The invasion of the unknown is of the nature of an adventure" (Dewey, 1930: 174).

References

Benjamin, J. (1998). *Shadow of the other: Intersubjectivity and gender in psychoanalysis.* London and New York: Routledge.

Bibby, T. (2009). How do children understand themselves as learners? Towards a learner-centred understanding of pedagogy, *Pedagogy, Culture & Society*, 17:1, 41–55.

Bibby, T. (2011). *Education – an "impossible profession"? Psychoanalytic explorations of learning and classrooms.* London and New York: Routledge.

Bion, W.R. (1962). *Learning from experience.* London: Karnac Books.

Bion, W.R. (2004) *Experiences in groups and other papers.* Hove: Brunner-
Routledge.

Bourdieu, P., & Passeron, J.C. (1977). *Reproduction in education, society and culture.* London: Sage.

Bourdieu, P. (1986). The forms of capital. In J.G. Richardson (Ed.), *Handbook of theory and research for the sociology of education.* New York: Greenwood Press.

Bourdieu, P. (1990). *The logic of practice.* Cambridge: Polity Press.

Bourdieu, P. (1993). *Sociology in question.* London: Sage.

Bourdieu, P. (1995). *Language and symbolic power.* Cambridge, MA: Harvard University Press.

Bourdieu, P. (1999). *Distinction: A social critique of the judgment of taste.* Cambridge, MA: Harvard University Press.

Bowlby, J. (1969). *Attachment and loss. Volume 1: Attachment.* London: Basic Books.

Britzman, D. (2003). *After-education: Anna Freud, Melanie Klein, and psychoanalytic histories of learning.* Albany: State University of New York Press.

Brooker, L. (2007). Changing the landscape of early childhood. In J. Moyles (Ed.), *Early years foundations. Meeting the challenge.* Maidenhead: Open University Press.

Chodorow, N. (1999). Toward a relational individualism. In Stephen A. Mitchell & Lewis Aron (Eds.), *Relational psychoanalysis: The emergence of a tradition.* New York and London: Routledge.

Connolly, P. (2000). Racism and young girls' peer-group relations: The experiences of South Asian girls. *Sociology*, 34(3): 499–519.

Connolly, P. (2004). *Boys and schooling in the early years.* London: Routledge Falmer.

Devine, D. (2009). Mobilising capitals? Migrant children's negotiation of their everyday lives in school. *British Journal of Sociology of Education*, 30(5): 521–535.

Durkheim, E. (1977). *The evolution of educational thought: Lectures on the formulation and development of secondary education in France*. Trans. by Peter Collins. London: Routledge and Kegan Paul.
Dewey, J. (1916). *Democracy and education*. New York: Free Press.
Dewey, J. (1930). *Democracy and education. An introduction to the philosophy of education*. New York: The MacMillan Company.
Dewey, J. (1938). *Experience and education*. New York: Macmillan Publishing.
Dimen M. (2003). *Sexuality, intimacy, power – relational perspectives*. London and New York: Routledge.
Dunn, J. (2004). *Children's friendships: The beginnings of intimacy*. Malden, MA: Blackwell.
Elias, N. (2007). *An essay on time*. Dublin: UCD Press [Collected Works, vol. 9].
Elias, N., & Dunning, E. (2008). *Quest for excitement*, vol. 7. Collected Works. Dublin: UCD Press.
Elias, N. (2009). On human beings and their emotions: A process-sociological essay. In *Essays III: On sociology and the humanities*. Dublin: UCD Press [Collected Works, vol. 16], pp. 141–158.
Elias, N. (2012). *On the process of civilisation*. Dublin: UCD Press [Collected Works, vol. 3].
Elkin, F., & Handel, G. (1972). *The child and society: The process of socialization*, 2nd edition. New York: Random House.
Farrell, A., Kagan, S.L., & Tisdall, E.K.M. (2015). *The SAGE handbook of early childhood research*. London: Sage.
Greenberg, J., & Mitchell, S. (1983). *Object-relations in psychoanalytic theory*. Cambridge, MA: Harvard University Press.
Hendrick, H. (1997). *Children, childhood and English society, 1880–1890*. Cambridge: Cambridge University Press.
James, A., & Prout, A. (Eds.) (1997). Introduction. In *Constructing and reconstructing childhood*, 2nd edition. London: Falmer.
James, A., Jenks, C., & Prout, A. (1998). *Theorizing childhood*. Cambridge: Polity.
Lister, R. (2006) Children (but not women) first: New labour child welfare and gender. *Critical Social Policy*, 26(2): 315–335.
Macmurray, J. (1957). *The self as agent*. London: Faber.
Macmurray, J. (1961). *Persons in relation: Vol. II of the Form of the personal*. London: Faber and Faber.
Macmurray, J. (1965). Reflections on the notion of an educated man. Invited paper, University of Bristol Institute of Education, 17 November, Unpublished.
Macmurray, J. (2012). Learning to be human. *Oxford Review of Education*, 38(6): 661–674.
Moss, P. (2014). *Transformative change and real utopias in early childhood education: A story of democracy, experimentation, and potentiality*. London: Routledge.
OECD. (2006). *Starting strong 2. Early childhood education and care*. Paris: OECD.
Palludan, C. (2007). Two tones: The core of inequality in kindergarten. *International Journal of Early Childhood*, 39(1): 75–91.
Papatheodorou, T. (2009). Exploring relational pedagogy, Chapter 1 in *Learning together in the early years: Exploring relational pedagogy*. London: Routledge.
Parsons, T. (1951). *The social system*. Glencoe: Free Press.
Parsons, T. (1959). The school class as a social system some of its functions in American society. *Harvard Educational Review*, 29: 297–318.
Petrie, P. (2003). Social pedagogy: An historical account of care and education as social control. In J. Brannen & P. Moss (Eds.), *Rethinking children's care*. Buckingham: Open University Press.
Prout, A. (2005). *The future of childhood*. London: Routledge/Falmer.
Prout, A. (2011). Taking a step away from modernity: Reconsidering the new sociology of childhood. *Global Studies of Childhood*, 1(1): 4–14.
Richards, M.P.R. (1974). Introduction. In M.P.R. Richards (Ed.), *The integration of a child into a social world*. Cambridge: Cambridge University Press, pp. 1–10.
Sahlberg, P. (2016). Professional autonomy, trust and collaboration in educators' work. Philosophy of Education Society of Great Britain. Annual Conference, New College, Oxford1–3 April 2016. http://www.philosophy-of-education.org/dotAsset/bc4f09b5-a27f-4306-9966-3aa1fa2cf12c.pdf
Quennerstedt, A., & Quennerstedt, M. (2014). Researching children's rights in education: sociology of childhood encountering educational theory. *British Journal of Sociology of Education*, 35:1, 115–132.

Robertson, S.L. (2015). What teachers need to know about the 'Global Education Reform Movement' (or GERM). In G. Little (Ed.), *Global education 'reform'. Building resistance and solidarity.* London: Manifesto Press, pp. 10–17.

Roseneil, S., & Ketokivi, K. (2016). Relational persons and relational processes: Developing the notion of relationality for the sociology of personal life. *Sociology*, 50(1): 143–159.

Sahlberg, P. (2014). *Finnish lessons 2.0: What can the world learn from educational change in Finland?*2nd edition. New York: Columbia University Press.

Statham, J., & Smith, M. (2010). *Issues in earlier intervention: Identifying and supporting children with additional needs.* London: HMSO: Department of Children, Schools and Families.

Sullivan, H.S. (1953). *The interpersonal theory of psychiatry.* New York: The William Alanson White Psychiatric Foundation.

Vuorisalo, M. (2011). Children's resources in action – the conversion of capital in the pre-school field. In L. Alanen, & M. Siisiäinen (Eds.), *Fields and capitals: Constructing local life.* Jyväskylä: University of Jyväskylä, Institute for Educational Research, pp. 29–60.

Wacquant, L. (2014). Homines in extremis: What fighting scholars teach us about habitus. *Body and Society*, 20(2): 3–17.

Winnicott, D.W. (1950). Growth and development in maturity. In D.W. Winnicott, *The family and individual development.* London: Tavistock.

Winnicott, D.W. (1964). *The child, the family, and the outside world.* London: Penguin.

Winnicott, D.W. (1965). *The family and individual development.* London: Tavistock Publications.

Winnicott, D.W. (1971). *Playing and reality.* London: Routledge.

9

THE CONTRIBUTIONS OF ECONOMICS TO EARLY CHILDHOOD EDUCATION AND CARE

W. Steven Barnett, Raquel Bernal and Milagros Nores

Introduction

The economic argument that early childhood education and care (ECEC) yields high rates of return has contributed to increased attention to public funding for ECEC in the United States and globally (Karoly, 2016; Nores & Barnett, 2010). ECEC has been integrated into the Sustainable Development Goals (Britto et al., 2017), as well as the G20 2018 communiqué (point 14). Traditionally, ECEC was viewed as primarily a private responsibility provided informally by families. As formal ECEC outside the home has become more common, in many countries public funding has increased. In the United States, most public spending at the federal level is through Head Start and child care subsidies for children in low-income families (Barnett & Kasmin, 2016). State and local government funding is concentrated on preschool education including special education.

Research on the economics of ECEC is most often thought of as benefit–cost analysis (BCA) designed to quantify the overall economic returns to public investments. As important as that topic is, economics has additional insights to offer regarding ECEC policy. For example, from the perspective of children and families, ECEC is not just an investment to be judged based on its future impacts on the economy; it is also a service that is valued because of its current contribution to the child's quality of life and because of its potential to reduce socioeconomic inequalities. Economics also offers insights into how ECEC produces its benefits, who benefits, and under what circumstances ECEC programs can be expected to be most beneficial.

Economic research on ECEC has built upon work in other disciplines on how ECEC programs directly and indirectly impact development as elements of the social environment and the notion of early childhood as a period in which the brain is especially responsive to environmental enrichment (Phillips & Shonkoff, 2000). Evidence that the under-resourced early environments of children in poverty offer a particularly important opportunity for public investments to improve their life courses has also been important (Barnett, 2011; Black et al., 2017; Cunha Heckman, Lochner, & Masterov, 2006; Barnett, 2011, Yoshikawa et al., 2016). Studies indicating that different pathways from ECEC support long-term cognitive benefits and social-emotional benefits (e.g., Reynolds, Ou, & Topitzes, 2004) have led economists to hypothesize domain specific life-cycle dynamics in the production of human capital (Heckman, Pinto, & Savelyev, 2013).

In this chapter, we review the advances in theory and applied research contributed by economics to the ECEC field, the economic rationale for public investments in ECEC, and the evidence regarding the economic returns to ECEC and the economic rationale for public funding.

The Human Capital Production Function

Human capital is the stock of an individual's capabilities and other characteristics that contribute to the production of goods and services very broadly defined to include not just the market economy but also family life and child rearing, political participation, self-care (e.g., health), and leisure (Becker, 1975). For this reason, human capital is a construct intertwined not just with an individual's productivity in the labor market, but with overall economic and social well-being or quality of life. Aggregate human capital accumulation is, thus, expected to determine economic growth and development, poverty alleviation, social mobility, and a way to break the intergenerational cycle of poverty and inequality.

For many years, much of the economics of education focused primarily on the human capital accumulated by investing in schooling and learning on-the-job and on its contribution to earnings (e.g., Ben-Porath, 1967; Griliches, 1977; Mincer, 1974). More recently, economists have devoted greater attention to early human capital accumulation and the role of the family and other aspects of the social and physical environment on the brain development and human functioning including the formation of a broad range of skills during the first years of life (Barnett, 1985, 1992, 2011; Becker, 1991; Heckman, 2000, 2008; Blau & Currie, 2006; Duncan & Brooks-Gunn, 1997).

Research on the human capital accumulation process over the life cycle found that socio-economic gaps in skill formation arise before entry to primary school, and subsequent schooling does not effectively close these gaps (Rubio-Codina et al., 2015). Cognitive and language gaps are already present by age 3 in a wide range of countries (Barnett & Lamy, 2013; Duncan & Magnuson, 2013; Fernald, Kariger, Hidrobo, & Gertler, 2012; Heckman, 2008; Rubio-Codina et al., 2015). Reducing or avoiding the emergence of these gaps early on would be easier and possibly more cost-effective than investing later over the life cycle. As a result, it became critical to understand better how early investments matter for early human capital formation.

One model of the early human capital production function characterizes its production as a dynamic process that depends on: the initial stock of skills (or genetic endowments), human capital investments at different stages of childhood, other parental and child characteristics, and random influences (Cunha & Heckman, 2007; Cunha, Heckman, & Schennach, 2010). To build children's human capital, parents and other care providers invest their time and purchase goods and services including toys, books, and nutrition and health services (Bernal, 2008; Bernal & Keane, 2010; Attanasio et al., 2018; Attanasio, Meghir & Nix, 2017). This approach recognizes inputs from outside the family including the quality of the learning environment in ECEC programs (Liu, Mroz & van der Klaauw, 2010).

Two methodological issues have particularly concerned economists studying the early human capital production function: possible endogeneity of investments and measurement error for both child development and parental investments. The production function is difficult to estimate correctly if investments are endogenous. For example, parents might decide how much to invest in a child depending on the unobserved (to the researcher) skills of the child or parent. If parents reinforce strengths or compensate weaknesses in child or parental abilities, it is difficult to disentangle the effects of investments on child development from preexisting differences in child and parental skills. To reduce address endogeneity bias, economists have adopted a variety of empirical approaches including structural estimation (Bernal, 2008; Liu et al., 2010), estimation by instrumental variables (Bernal & Keane, 2010; Attanasio et al., 2018), and correction by control function (Attanasio et al., 2019). Accounting for endogeneity tends to increase estimated effects of investments in early development (Attanasio et al., 2018; Cunha et al., 2010; Attanasio et al., 2017; Liu et al., 2010).

Studies estimating early human capital production functions typically use a variety of measures of learning and development as well as measures of parental investments that contain significant measurement error. In this literature, it is common to implement a latent factor model approach to summarize the information contained in the different developmental measures in a smaller number of interpretable constructs. This facilitates correction for measurement error contained in the measures. Specifically, researchers specify a measurement system in which each measure is associated with at most one developmental or learning construct (Heckman et al., 2013). Once the parameters of the measurement system have been estimated, these are used to estimate factor scores (Cunha et al., 2010).

By providing consistent estimates of the early human capital production function, economic research has contributed to three aspects of the understanding of early childhood development and the investments that can foster it. These are: how various investments benefit child development; how these benefits may vary with the characteristics of the children and families involved; and, the design of more efficient public investments (getting a bigger bang for the buck).

Production function research has sought to identify the *mechanisms* through which investments affect child development and how the *constraints* faced by families may limit these investments. For example, Attanasio et al. (2018) estimated a human capital production function to investigate mechanisms through which a home visitation program impacted the development of socioeconomically vulnerable children in Colombia. They found that the primary driver of program impacts was an increase in time and materials provided by parents rather an improvement in parenting skills. Cunha, Elo, and Culhane (2013) found that mothers underestimate the productivity of their investments in children by about a half, which leads them to underinvest about a fifth less than if they had known their true productivity. These studies suggest that research and practice should focus the quantity as well as quality of parent–child interactions.

Another stream of research has sought to understand who benefits the most from various public and private investments in ECEC (e.g., Brilli, Del Boca, & Pronzato, 2016; Felfe & Lalive, 2018; Kottelenberg & Lehrer, 2017). Results appear to depend on both the features of the ECEC and the broader context, including labor market conditions. Generally, economists find that better educated parents invest more of their own time in each child (Guryan, Hurst, & Kearney, 2008). Attanasio et al. (2018) report that parents invest more in children with higher initial cognitive skills. Given the very early emergence of socioeconomic developmental gaps, a case can be made to target public investments on the most vulnerable children as wealthier families are already investing more (e.g., Kottelenberg & Lehrer, 2017). However, there are no easy answers as some research finds universal programs may be more effective than targeted programs for disadvantaged children (Barnett, 2010; Cascio, 2017).

Economic research also has sought insights to improve the *efficiency* of interventions, for example, regarding the returns to starting earlier or providing a larger "dosage." Cunha et al. (2010) found that it is more effective to invest earlier than later during childhood. Attanasio et al. (2019), using rich high-frequency data for children 2 to 7 years of age in a randomized controlled trial that assigned children to a high-quality center-based childcare program in Colombia, found that ages 3 and 4 are a more productive period to target interventions aimed at improving parental investments. Yet much remains to be learned about the dynamics of early skill formation, the interactions among different inputs relating to nutrition, health, and learning, and optimal timing and duration of various interventions, among other things.

Rationales for Public Investment in ECEC

Economics has put forward three primary rationales for public investments in ECEC: equity and redistribution, market failures, and economic returns. Together these support the view that ECEC that enables more children to reach their full developmental potential is a public good worthy of government investment. We consider each rationale in turn below, but first we briefly review research linking the immediate impacts of ECEC on development to long-term economic outcomes. This research critically informs benefit–cost analyses that seek to estimate the economic returns to ECEC programs.

Economic studies have linked early childhood development to long-term benefits to individuals and to societal (macroeconomic) outcomes including economic growth and development, decreased inequality, and intergenerational social mobility (Bivens et al., 2016). Research links early cognitive abilities to later school achievement, educational attainment, employment, and earnings (e.g., Currie & Almond, 2011; Bernal & Keane, 2010). Similarly, emotional and behavioral problems measured at age 6 to 8 predict high school and college graduation (McLeod & Kaiser, 2004). Currie and Rossin-Slater (2015) summarize an extensive literature linking early-life health measures including prenatal nutrition to later cognitive abilities, educational attainment, adult employment, and earnings as well as adult health. This research allows extrapolation from ECEC impacts on early childhood development and health to adult well-being and benefits to society, including economic growth and development, even when only short-term outcomes are measured.

Equity

Differences in early parental investments in learning and development in the home and in formal ECEC exacerbate inequality from an early age (Conti & Heckman, 2013; Schady et al., 2015). Public investment in ECEC offers a way to promote increased equality (Britto, Yoshikawa, & Boller, 2011). Public investment can offset advantages due to parental education and income, but also provide greater support for children whose parents do not prioritize their development and for inequities related to child characteristics. For example, some parents might invest more in boys than girls, in the first born, or based on other child characteristics. Direct provision of early childhood services, as an alternative to cash transfers or loans, may be needed to ensure that all children are provided the same opportunities regardless of their parents' preferences. This equity perspective also aligns with a view that children are citizens with their own rights, which government is obligated to protect.

Market Failures

Market failures result from information problems, liquidity constraints (and credit market failures), and/or externalities. Information problems can arise if parents have difficulties correctly assessing the quality of ECEC or its impacts on child development that might lead them to underinvest in ECEC (Bassok, Markowitz, Player, & Zagardo, 2017). Liquidity constraints arise because the cost of high-quality ECEC is high relative to the incomes of many families with young children, and borrowing to finance ECEC is limited by the uncertainty of returns to individuals and the many years required for benefits to accrue. Externalities occur when ECEC produces benefits beyond those that accrue to the individual participants. As discussed below, economic research finds extensive, large externalities. Information problems can be addressed by initiatives that seek to better inform parents or regulate ECEC. Liquidity problems can be addressed by a range of government interventions. Externalities may require public subsidies and possibly public purchase or provision of ECEC to ensure that societal benefits are fully maximized if private and public interests diverge.

Economic Returns: Benefit-Cost Analysis (BCA)

Studies on the returns to ECEC weigh its benefits against its costs. This requires an accurate estimation of all of the costs of attending a program, public and private, and as many of its benefits as can be measured and valued. As ECEC produces a service that can jointly benefit parents (child care) as well as children (education and development), BCAs should consider both the earnings benefits to parents from increased work hours and the wide range of potential benefits to children. The children's benefits most commonly included are lower costs of schooling (less grade repetition and special education), higher earnings as adults from increased educational achievement and attainment, reduced costs of crime, and the value of better health (Karoly, 2016). These benefits can generate substantial externalities including decreased costs to government for public education, the criminal justice system, health care, and social services (welfare and child abuse) as well as higher tax revenues and less crime victimization. BCA results typically are summarized as rates of return or benefit–cost (B/C) ratios, which indicate a positive return if greater than 1.

The Evidence:Experimental and Non-Experimental Studies

BCA depends on estimates of ECEC effects, and there are many studies of ECEC effects that could support BCAs. However, BCAs of ECEC are relatively uncommon and not necessarily representative of the larger literature. BCAs tend to be conducted for studies with more positive results. This creates a kind of "file drawer" problem as the least effective ECEC policies and programs are not subject to BCA. One meta-analysis found that 109 of 135 early intervention studies found at least one positive impact, and 29 percent of estimated effects were positive overall (Cannon et al., 2018). One approach to dealing with this issue is to conduct BCAs based on meta-analysis results as well as for individual programs or policies.

In the United States, studies of ECEC tend to find positive immediate effects (Camilli, Vargas, Ryan, & Barnett, 2010; Duncan & Magnuson, 2013). However, effects typically shrink after children enter primary school. In some studies, some substantive effects persist on cognitive abilities including achievement despite shrinkage. In other studies cognitive effects disappear but effects in other domains appear to persist that may produce economic benefits including increased earnings (Yoshikawa et al., 2013). The size and persistence of effects varies with the population served, the political, social and economic contexts, and program characteristics.

Overall, the evidence on ECEC impacts globally – including in low- and middle-income countries – is similar in its findings to that in the United States (Britto et al., 2017; Nores & Barnett, 2010; Rao, Sun, Chen, & Ip, 2017). Although effects tend to decline in magnitude after children exit ECEC, positive effects can persist through the school years and into adulthood. Again, there is substantial variation in outcomes. Estimated effects on children vary strongly with the quality and mix of services (e.g., health, nutrition, education, care) provided by ECEC programs.

Three Seminal Studies

The Perry Preschool study, which began in 1962, is among the most influential in the ECEC field. It followed a sample of 123 low-income children from ages 3 and 4 through age 40 (Schweinhart et al., 2005). Children were randomly assigned to treatment and control groups. The treatment group received 2.5 hours per day of high-quality preschool and weekly home-visits for up to 2 years. Positive effects on cognitive test scores were observed from ages 3 through 27, but persistent effects were found only for achievement and related outcomes, not for IQ. Both self-report and official records data indicated that those who had attended the preschool were less involved with delinquency and crime. Other positive long-term outcomes included fewer years of special education years, and a higher rate of on-time high school graduation. Positive adult outcomes include increased earnings, decreased welfare dependency, decreases in teen pregnancy, and, in association with the crime reductions, fewer arrests.

BCAs of the Perry Preschool program and its outcomes were conducted periodically as the study progressed. The earliest BCAs depended heavily on projections based on the other studies' estimates of links between early and later outcomes. As the study progressed projections were increasingly replaced by estimates based on observed outcomes. By age 40, the estimated benefits from observed outcomes included educational cost reductions, delinquency and crime victim and criminal justice system cost reductions, welfare cost reductions, and earnings (Nores, Belfield, Barnett, & Schweinhart, 2005).

BCA estimates of the economic return from the Perry Preschool varied over the years primarily because of the shift from projected to observed outcomes. All the estimates have been much higher than breakeven, with the highest 14 to 1. Recent re-analyses by Heckman and colleagues relied on substantially different assumptions about the value of benefits that lowered the estimated rate of return to about 9 to 1 as reported in Table 9.1 (Heckman, Moon, Pinto, Savelyev, & Yavitz, 2010).

Two important insights were evident from even the earliest of the BCAs of the Perry Preschool program and its outcomes. First, a large part of the economic benefit was from impacts on outcomes other than earnings, including crime reduction. Second, most of the economic benefits were externalities that accrued to the general public rather than individual participants.

The Abecedarian study has followed 111 children from disadvantaged families since the 1970s through age 30 (Campbell et al., 2012). Children were randomly assigned to an ECEC program or control group from the first months of life. The ECEC consisted of center-based child care with an emphasis on language development, eight hours a day, year round, through age 5. Positive effects have been found for IQ and reading and math achievement in the short- and long-term. Children who attended the program had much lower rates of grade retention and special education during the school years and higher rates of participation in higher education. Positive effects were also found for health-related behaviors and symptoms of depression (Campbell et al., 2012).

Benefits in the Abecedarian study also exceeded costs with a benefit–cost ratio of 2.5:1 based on data through age 21 (Table 9.1; Barnett & Masse, 2007). Externalities were considerable. However, the estimated net benefits were lower than in the Perry study due to the higher costs of a full-day, year-round program for five years and lower benefit estimates. Also, no significant impacts were found on crime and delinquency. Evidence regarding health benefits was reported after the initial benefit–cost analysis was conducted (Muennig et al., 2011). Estimates of benefits from maternal earnings were difficult to estimate due to limitations of early data collections so that they may not be fully included in the analyses either. For these reasons, the estimate in Table 9.1 likely underestimates the return.

The Chicago Child–Parent Center (CPC) began during the 1980s, and consisted of health and social services, free meals, and center-based half-day preschool for 3- and 4-year-olds during the school year (Reynolds, Temple, White, Ou, & Robertson, 2011). The treated sample was matched with similar children and the overall sample is roughly ten times the size of the samples in the Perry and Abecedarian studies. As the CPC study is not a randomized trial, it is possible that unknown differences between treatment and comparisons introduce bias. However, this study has greater statistical power to detect modest effects and a more generalizable program design, population, and context than the other two seminal studies. It estimates the benefits of an ECEC program that might be replicable by today's public preschool programs at scale.

Despite differences from the two earlier studies, the pattern of effects and benefits is remarkably similar but the effect and benefit estimates tend to be smaller in CPC, suggesting a dose–response relationship. The CPC study found persistent increases in test scores, lower grade repetition and special education rates, higher high school completion rates, lower arrest rates, less use of food stamps, less substance abuse, and fewer cases of abuse and neglect. Reynolds et al. (2011) estimate that the benefit–cost ratio exceeded 10:1.

These three long-term studies and their costs, benefits, and rates of economic returns are displayed in Table 9.1. Dollar values are at present values at program entry (i.e., discounted at a rate of 3 percent annually to account for opportunity costs of waiting for the return), and adjusted for inflation to 2014 dollars for comparison. All of them are considerably greater than one, despite being incomplete. Even if only earnings benefits are considered, all the programs still pay off, though clearly much of the economic return is missed if only earnings are considered. Although it is tempting to make comparisons across the studies, all the estimates are highly uncertain due to large statistical confidence intervals and reliance on a variety of imperfect assumptions. Moreover, the key question for policy and practice is not what the precise rates of return were to these programs in their time and place, but what similar ECEC might yield at scale today.

TABLE 9.1 Features and returns per child (2014$) of the Perry Preschool, Abecedarian, and Chicago Child-Parent Center (CPC) programs

	Perry Preschool (n=123)	*Abecedarian (n=111)*	*Chicago CPC (n=1400)*
Ages Served	3 to 4	Six weeks to 5	3 to 4
Schedule	Half-day, school year	Full-day, year round	Half-day, school year
Earnings Benefits	$91,606	$147,359	$32,933
All Benefits	$179,446	$208,283	$105,204
Cost	$20,854	$85,530	$9,719
Returns Earnings Alone	*4.4 to 1*	*1.8 to 1*	*3.4 to 1*
Total Returns	*8.6 to 1*	*2.5 to 1*	*10.8 to 1*

Source: Ramon et al., 2018.

BCA of ECEC at Scale

As valuable as the three seminal studies have been, they tell us little about returns to large-scale publicly funded ECEC today, which sometimes have produced disappointing results. Some high-profile rigorous studies in the United States and Canada have found long-term impacts ranging from near zero to small negative effects (Barnett, 2011; Haeck, Lebihan, & Merrigan, 2018; Yoshikawa et al., 2016). Other studies have yielded more positive outcomes. This has spurred research on the features of programs that support stronger, more persistent positive effects and how results vary with the populations served and contexts. Studies find that ECEC quality matters, disadvantaged children benefit more, and results depend on the available alternative environments at home or in other ECEC programs (Haeck et al., 2018; Kottelenberg & Lehrer, 2017; Yoshikawa et al., 2016).

In view of the mixed results for child outcomes, it is unsurprising that the estimated economic returns to large-scale public programs are highly varied (Karoly, 2016). Early Head Start is estimated to return less than its cost. Some have concluded that Head Start is unlikely to have a positive return, but others estimate B/C ratios of $3 to 1 based on small positive long-term effects that are consistent with the confidence intervals estimated in the national randomized trial. Other public preschool programs supported by state and local governments vary greatly in their design and outcomes. Kay and Pennucci (2014) estimated about a $2 to 1 B/C ratio based on average effects from a meta-analysis. Estimates for specific state programs range from $2 to $4 to 1 and even higher (Bartik, Gormley, & Adelstein, 2012; Cascio & Schanzenbach, 2013; Karoly, 2016; Ramon et al., 2018). These BCAs heavily rely on projections from estimated impacts on achievement test scores for future earnings projections.

Beyond the United States, findings regarding the impacts and benefits of public preschool are at least as mixed (Nores & Barnett, 2010; Rao et al., 2017; van Huizen & Plantenga, 2018). Again, quality of ECEC matters for effectiveness, as does the population served and context (Britto et al., 2017). Despite a substantial research literature, BCAs of ECEC outside North America are rare. The UK's Sure Start local programs were found to have a modest positive impact on parental earnings, but not enough for a positive economic return, and benefits to child development could not be estimated (Dalziel, Halliday, & Segal, 2015). A BCA of lowering the eligibility age for universal preschool to age 3 estimated a return above $4 to 1 with earnings benefits from enhanced learning and development dominating the returns though benefits from increased maternal employment were included (van Huizen, Dumhs, & Plantenga, 2017).

The lack of BCAs in middle- and low-income countries is particularly notable given the extensive research on ECEC impacts in these countries (Nores & Barnett, 2010; Rao et al., 2017; Britto et al., 2017). BCAs of ECEC have been conducted in Latin America, relying on projected earnings benefits due to improvements in learning and development with returns ranging from just over $1.1 to $5.1 (Armendáriz, Ardanaz, Behrman, Cristia, & Hincapie, 2015). Engle et al. (2011) estimated that increasing preschool education enrollments in low- to middle-income countries from 25 to 50 percent could yield a much higher benefit-to-cost ratio of between 6 and 17 to 1 based on increased earnings alone.

Economists have studied extensively the effects of ECEC on maternal employment and earnings, but these benefits are less often incorporated into comprehensive BCA of large-scale programs (Blau & Currie, 2006; Vuri, 2016). Parental earnings benefits vary with ECEC eligibility criteria, ECEC operating hours, other policies supporting maternal employment, labor market conditions, rates of single parenting, and other determinants of maternal employment. In the United States, one study suggests that the Head Start program decreased maternal labor supply (Long, 2016), and Fitzpatrick (2010) failed to find effects on maternal labor supply of universal pre-K in Georgia or Oklahoma. Research in other countries has found everything from null effects, to small positive effects for some subgroups to substantial positive average effects (Brewer & Cattan, 2017). Quebec's universal child care program had relatively large estimated impacts but researchers differ regarding the economic benefit relative to cost (Fortin, Godbout, & St-Cerny, 2012; Haeck, Lefebvre, & Merrigan, 2015).

Limitations and Next Steps

Research on the economics of ECEC has provided strong evidence that ECEC can be a sound public investment based on the magnitude of its benefits and the extent to which these are externalities. However, large-scale public investments in ECEC have not always produced the hoped for benefits. If ECEC investments are to fulfil their promise, policy makers must attend to the details that influence who is served and what is delivered. In most contexts, child benefits from education and related services (e.g., health and nutrition) will be much more important for securing an adequate return than impacts on parental employment. Children who can benefit most must be included among the recipients, and the quality of services must be high. Beyond this, more remains to be learned about how to optimize economic returns to public investments in ECEC.

As benefits depend greatly on who is served, program implementation, and social context, the incorporation data systems for continuous improvement is needed for ECEC success at scale (Barnett & Frede, 2017; Nores & Fernandez, 2018). Intensity, duration, and quality of ECEC matter for specific populations at different ages, but in ways that may be difficult to generalize. For example, the potential returns to increasing quality are large, but what this means likely varies considerably across countries (Engle et al., 2011).

For benefit–cost analysis to be useful to policy makers, it must be more widely applied, and economists must improve on estimating cost and benefits (Dalziel et al., 2015). Karoly (2012) sets out in detail a standardized methodology for pursuing such work. As discussed earlier, child care benefits to parents should be included but are unlikely to be enough to justify public investments in ECEC, which warrants more emphasis on a full accounting of benefits for children. Early test scores provide a risky basis for predicting outcomes so it would be better not to rely solely on these whether for earnings or such other benefits as crime reduction, health care cost savings, or longer, healthier lives. Finally, broader societal impacts should be considered, including social inequality and social cohesion, even if it is not possible to put a dollar value on them. As a field, economic analysis of ECEC is still quite young. Investments in its healthy development might well produce its own impressive long-term benefits.

References

Armendáriz, E., Ardanaz, M., Behrmen, J.R., Christia, J., & Hincaple, D. (2015). More bang for the buck: investing in early childhood development. In S. Berlinski & N. Schady (Eds.), *The early years: child well-being and the role of public policy* (pp. 149–178). Basingstoke: The Inter-American Development Bank, Palgrave Macmillan.

Attanasio, O., Meghir, C., & Nix, E. (2017). Investments in children and the development of cognition and health in India. NBER Working Paper No. 21740.

Attanasio, O., Bernal, R., Giannola, M., & Nores, M. (2019). Child development in the early years: Parental investments and the changing dynamics of different dimensions. Unpublished manuscript University College London, Rutgers University and Universidad de los Andes.

Attanasio, O., Cattan, S., Fitzsimons, E., Meghir, C., & Rubio-Codina, M. (2018). Estimating the production function for human capital: results from a randomized control trial in Colombia. NBER Working Paper No. 20965.
Barnett, W.S. (1985). Benefit–cost analysis of the Perry Preschool Program and its policy implications. *Educational Evaluation and Policy Analysis*, 7(4), 333–342.
Barnett, W.S. (1992). Benefits of compensatory preschool education. *Journal of Human Resources*, 27(2), 279–312.
Barnett, W.S. (2010). Universal and targeted approaches to preschool education in the United States. *International Journal of Child Care and Education Policy*, 4(1), 1.
Barnett, W.S. (2011). Effectiveness of early educational intervention. *Science*, 333(6045), 975–978.
Barnett, W.S., & Frede, E.C. (2017). Long-term effects of a system of high-quality universal preschool education. In H.P. Blossfeld, N. Kulic, J. Skopek, and M. Triventi (Eds.), *Childcare, early education, and social inequality: An international perspective* (ch. 8). Cheltenham, UK and Northampton, MA: Edward Elgar Publishing. doi:10.4337/9781786432094.00018
Barnett, W.S., & Kasmin, R. (2016). Funding landscape for preschool with a highly qualified workforce. Paper prepared for the Board on Children, Youth, and Families of the National Academy of Sciences, Engineering, and Medicine. http://sites.nationalacademies.org/cs/groups/dbassesite/documents/webpage/dbasse_176099.pdf
Barnett, W.S., & Lamy, C.E. (2013). Achievement gaps start early. In P. Carter & K. Welner (Eds.), *Closing the opportunity gap: What America must do to give every child an even chance* (pp. 98–110). Oxford: Oxford University Press.
Barnett, W.S., & Masse, L.N. (2007). Comparative benefit–cost analysis of the Abecedarian program and its policy implications. *Economics of Education Review*, 26(1), 113–125.
Bartik, T.J., Gormley, W., & Adelstein, S. (2012). Earnings benefits of Tulsa's pre-K program for different income groups. *Economics of Education Review*, 31(6), 1143–1161.
Bassok, D., Markowitz, A.J., Player, D., & Zagardo, M. (2017). Do parents know "high quality" preschool when they see it? EdPolicyWorks Working Paper Series No. 54. Charlottesville, VA: University of Virginia Curry School.
Becker, G.S. (1975). *Human capital: A theoretical and empirical analysis, with special reference to education* (2nd ed.). New York: Columbia University Press for NBER.
Becker, G.S. (1991). *A treatise on the family*. Cambridge, Mass: Harvard University Press.
Ben-Porath, Y. (1967). The production of human capital and the life cycle of earnings. *Journal of Political Economy*, 75(4 Part 1), 352–365.
Bernal, R. (2008). The effect of maternal employment and child care on children's cognitive development. *International Economic Review*, 49(4), 1173–1209.
Bernal, R., & Keane, M.P. (2010). Quasi-structural estimation of a model of child Care choices and cognitive ability production. *Journal of Econometrics*, 156(1), 164–189.
Bivens, J., Garcia, E., Gould, E., Weiss, E., & Wilson, V. (2016). It's time for an ambitious national investment in America's children: Investments in early childhood care and education would have enormous benefits for children, families, society, and the economy. Washington, DC: Economic Policy Institute. https://www.epi.org/publication/its-time-for-an-ambitious-national-investment-in-americas-children/
Black, M.M., Walker, S.P., Fernald, L.C., Andersen, C.T., DiGirolamo, A.M., Lu, C., … & Devercelli, A.E. (2017). Early childhood development coming of age: science through the life course. *The Lancet*, 389(10064), 77–90.
Blau, D., & Currie, J. (2006). Pre-school, day care, and after-school care: who's minding the kids? In E. Hanushek & F. Welch (Eds.), *Handbook of the economics of education* (pp. 1163–1276). Amsterdam: North-Holland.
Brewer, M., & Cattan, S. (2017). Universal pre-school and labor supply of mothers. *ifo DICE Report*, 15 (2), 8–12.
Brilli, Y., Del Boca, D., & Pronzato, C.D. (2016). Does child care availability play a role in maternal employment and children's development? Evidence from Italy. *Review of Economics of the Household*, 14(1), 27–51.

Britto, P.R., Yoshikawa, H., & Boller, K. (2011). Social policy report quality of early childhood development programs in global contexts rationale for investment, conceptual framework and implications for equity. *Social Policy Report*, 25(2).
Britto, P.R., Lye, S.J., Proulx, K., Yousafzai, A.K., Matthews, S.G., Vaivada, T., ... & MacMillan, H. (2017). Nurturing care: promoting early childhood development. *The Lancet*, 389(10064), 91–102.
Camilli, G., Vargas, S., Ryan, S., & Barnett, W.S. (2010). Meta-analysis of the effects of early education interventions on cognitive and social development. *The Teachers College Record*, 112(3). Retrieved from http://www.tcrecord.org/Content.asp?ContentID=15440
Campbell, F.A., Pungello, E.P., Burchinal, M., Kainz, K., Pan, Y., Wasik, B.H., ... & Ramey, C.T. (2012). Adult outcomes as a function of an early childhood educational program: an Abecedarian Project follow-up. *Developmental Psychology*, 48(4), 1033.
Cannon, J.S., Kilburn, M.R., Karoly, L.A., Mattox, T., Muchow, A.N., & Buenaventura, M. (2018). Investing early: taking stock of outcomes and economic returns from early childhood programs. *Rand Health Quarterly*, 7(4), 6.
Cascio, E.U. (2017). Does universal preschool hit the target? Program access and preschool impacts (No. w23215). National Bureau of Economic Research.
Cascio, E.U., & Schanzenbach, D.W. (2013). The impacts of expanding access to high-quality preschool education (No. w19735). National Bureau of Economic Research.
Conti, G. & Heckman, J.J. (2013). The developmental approach to child and adult health. *Pediatrics*, 131(Supplement), S133–S141.
Cunha, F., & Heckman, J. (2007). The technology of skill formation. *American Economic Review*, 97(2), 31–47.
Cunha, F., Elo, I., & Culhane, J. (2013). Eliciting maternal expectations about the technology of cognitive skills formation (No. w19144). National Bureau of Economic Research.
Cunha, F., Heckman, J., & Schennach, S. (2010). Estimating the technology of cognitive and noncognitive skill formation. *Econometrica*, 78(3), 883–931.
Cunha F., Heckman J., Lochner, L., & Masterov, D. (2006). Interpreting the evidence on life cycle skill formation. In E. Hanushek & F. Welch (Eds.), *Handbook of the economics of education* (vol. 1, pp. 697–812). Amsterdam: Elsevier.
Currie, J., & Almond, D. (2011). Human capital development before age five. In O. Ashenfelter, & D. Card (Eds.), *Handbook of labor economics* (vol. 4, Part B, pp. 1315–1486). Amsterdam: Elsevier.
Currie, J., & Rossin-Slater, M. (2015). Early-life origins of life-cycle well-being: research and policy implications. *Journal of Policy Analysis and Management*, 34(1), 208–242.
Dalziel, K.M., Halliday, D., & Segal, L. (2015). Assessment of the cost–benefit literature on early childhood education for vulnerable children: what the findings mean for policy. *Sage Open*, 5(1), doi:2158244015571637
Duncan, G., & Brooks-Gunn, J. (1997). Income effects across the life-span: Integration and interpretation. In G. Duncan & J. Brooks-Gunn (Eds.), *Consequences of growing up poor* (pp. 596–610). New York: Russell Sage Foundation.
Duncan, G., & Magnuson, K. (2013). Investing in preschool programs. *Journal of Economic Perspectives*, 27 (2), 109–132.
Engle, P.L., Fernald, L.C., Alderman, H., Behrman, J., O'Gara, C., Yousafzai, A., ... & Iltus, S. (2011). Strategies for reducing inequalities and improving developmental outcomes for young children in low-income and middle-income countries. *The Lancet*, 378(9799), 1339–1353.
Felfe, C., & Lalive, R. (2018). Does early child care affect children's development? *Journal of Public Economics*, 159, 33–53.
Fernald, L.Kariger, P., Hidrobo, M., & Gertler, P. (2012). Socioeconomic gradients in child development in very young children: evidence from India, Indonesia, Peru, and Senegal. *Proceedings of the National Academy of Sciences*, 109 (Supplement 2), 17273–17280.
Figlio, D., Guryan, J., Karbownik, K., & Roth, J. (2014). The effects of poor neonatal health on children's cognitive development. *American Economic Review*, 104(12), 3921–3955.
Fitzpatrick, M.D. (2010). Preschoolers enrolled and mothers at work? The effects of universal prekindergarten. *Journal of Labor Economics*, 28(1), 51–85.

Fortin, P., Godbout, L., & St-Cerny, S. (2012). Impact of Quebec's universal low fee childcare program on female labour force participation, domestic income, and government budgets (Working Paper 2012/02). Sherbrooke, Quebec: University of Sherbrooke.
Griliches, Z. (1977). Estimating the returns to schooling: Some econometric problems. *Econometrica*, 45 (1), 1–22.
Guryan, J., Hurst, E., & Kearney, M. (2008). Parental education and parental time with children. *Journal of Economic Perspectives*, 22(3), 23–46.
Haeck, C., Lebihan, L., & Merrigan, P. (2018). Universal child care and long-term effects on child well-being: evidence from Canada. *Journal of Human Capital*, 12(1), 38–98.
Haeck, C., Lefebvre, P., & Merrigan, P. (2015). Canadian evidence on ten years of universal preschool policies: the good and the bad. *Labour Economics*, 36, 137–157.
Heckman, J. (2000). Policies to foster human capital. *Research in Economics*, 4(51), 3–56.
Heckman, J. (2008). Schools, skills, and synapses. *Economic Inquiry*, 46(3), 289–324.
Heckman J., Pinto R., & Savelyev, P. (2013). Understanding the mechanisms through which an influential early childhood program boosted adult outcomes. *American Economic Review*, 103(6), 2052–2086.
Heckman, J.J., Moon, S.H., Pinto, R., Savelyev, P.A., & Yavitz, A. (2010). The rate of return to the HighScope Perry Preschool Program. *Journal of Public Economics*, 94(1–2), 114–128.
Karoly, L.A. (2012). Toward standardization of benefit-cost analysis of early childhood interventions. *Journal of Benefit-Cost Analysis*, 3(1), 1–45.
Karoly, L.A. (2016). The economic returns to early childhood education. *The Future of Children*, 26(2), 37–55.
Kay, N., & Pennucci, A. (2014). Early childhood education for low-income students: A review of the evidence and benefit-cost analysis. Olympia, WA: Washington State Institute for Public Policy.
Kottelenberg, M.J., & Lehrer, S.F. (2017). Targeted or universal coverage? Assessing heterogeneity in the effects of universal child care. *Journal of Labor Economics*, 35(3), 609–653.
Liu, H., Mroz, R., & van der Klaauw, W. (2010). Maternal employment, migration and child development. *Journal of Econometrics*, 156(1), 212–228.
Long, C. (2016). Introduction of Head Start and Maternal Labor Supply: Evidence from a Regression Discontinuity Design. US Census Bureau Center for Economic Studies Paper No. CES-WP-16-35.
McLeod, J., & Kaiser, K. (2004). Childhood emotional and behavioral problems and educational attainment. *American Sociological Review*, 69(5), 636–658.
Mincer, J. (1974). *Schooling, experience, and earnings*. New York: National Bureau of Economic Research, Columbia University Press.
Muennig, P., Robertson, D., Johnson, G., Campbell, F., Pungello, E.P., & Neidell, M. (2011). The effect of an early education program on adult health: the Carolina Abecedarian Project randomized controlled trial. *American Journal of Public Health*, 101(3), 512–516.
Nores, M., & Barnett, W.S. (2010). Benefits of early childhood interventions across the world: (Under) Investing in the very young. *Economics of Education Review*, 29(2), 271–282.
Nores, M., & Fernandez, C. (2018). Building capacity in health and education systems to deliver interventions that strengthen early child development. *Annals of the New York Academy of Sciences*, 1419(1), 57–73.
Nores, M., Belfield, C.R., Barnett, W.S., & Schweinhart, L. (2005). Updating the economic impacts of the High/Scope Perry Preschool program. *Educational Evaluation and Policy Analysis*, 27(3), 245–261.
Phillips, D.A., & Shonkoff, J.P. (Eds.). (2000). *From neurons to neighborhoods: The science of early childhood development*. Washington, DC: National Academies Press.
Ramon, I., Chattopadhyay, S.K., Barnett, W.S., & Hahn, R.A. (2018). Early childhood education to promote health equity: A Community Guide economic review. *Journal of Public Health Management and Practice*, 24(1), e8–e15. doi:10.1097/PHH.0000000000000557
Rao, N., Sun, J., Chen, E.E., & Ip, P. (2017). Effectiveness of early childhood interventions in promoting cognitive development in developing countries: A systematic review and meta-analysis. *Hong Kong Journal of Paediatrics*, 22(1), 14–25.
Reynolds, A.J., Ou, S.R., & Topitzes, J.W. (2004). Paths of effects of early childhood intervention on educational attainment and delinquency: A confirmatory analysis of the Chicago Child-Parent Centers. *Child Development*, 75(5), 1299–1328.

Reynolds, A.J., Temple, J.A., White, B.A., Ou, S.R., & Robertson, D.L. (2011). Age 26 cost–benefit analysis of the child-parent center early education program. *Child Development*, 82(1), 379–404.

Rubio-Codina, M., Attanasio, O., Meghir, C., Varela, N., & Grantham-McGregor, S. (2015). The socioeconomic gradient of child development: cross-sectional evidence from children 6–42 months in Bogotá. *Journal of Human Resources*, 50(2), 464–483.

Schady, N., Behrman, J., Araujo, M., Azuero, R., Bernal, R., Bravo, D., Lopez-Boo, F., Macours, K., Marshall, D., Paxson, C., & Vakis, R. (2015). Wealth gradients in early childhood cognitive development in five Latin American countries. *Journal of Human Resources*, 50(2), 446–463.

Schweinhart, L.J., Montie, J., Xiang, Z., Barnett, W.S., Belfield, C.R., & Nores, M. (2005). *Lifetime effects: The High/Scope Perry Preschool study through age 40.* Ypsilanti, MI: High/Scope Press.

van Huizen, T., & Plantenga, J. (2018). Do children benefit from universal early childhood education and care? A meta-analysis of evidence from natural experiments. *Economics of Education Review*, 66, 206–222.

van Huizen, T., Dumhs, L., & Plantenga, J. (2017). The costs and benefits of investing in universal preschool: Evidence from a Spanish reform. *Child Development*, 90(3), e386–e406.

Vuri, D. (2016). Do childcare policies increase maternal employment? *IZA World of Labor*. https://wol.iza.org/articles/do-childcare-policies-increase-maternal-employment/long

Yoshikawa, H., Weiland, C., & Brooks-Gunn, J. (2016). When does preschool matter? *The Future of Children*, 26(2), 21–35.

Yoshikawa, H., Weiland, C., Brooks-Gunn, J., Burchinal, M.R., Espinosa, L.M., Gormley, W.T., ... & Zaslow, M.J. (2013). Investing in our future: The evidence base on preschool education. https://www.srcd.org/policy-media/policy-updates/meetings-briefings/investing-our-future-evidence-base-preschool

10

THE IMPORTANCE OF THE PEDAGOGICAL SCIENCES TO EARLY CHILDHOOD EDUCATION

Sharon Ryan and M. Elizabeth Graue

Pedagogy or the practices and principles of teaching young children has been called "the silent partner" (Siraj-Blatchford, 1999; Stephen, 2010) in discussions about early childhood education. Yet it is impossible for curriculum, standards, policies etc. to be enacted without the teacher and the knowledge, skills, and dispositions, among other elements that s/he/they bring into daily classroom interactions. In this chapter we argue that the research base on pedagogical sciences in early childhood education is an important source of information for understanding and improving the quality and outcomes of children's experiences in early care and education settings. We contend that the study of pedagogy is a science in its own right that must be attended to if the field is to move forward.

To understand why pedagogy or teaching is often not at the forefront of discussions about early childhood education, this chapter begins with defining pedagogy, and early childhood pedagogy in particular, through an exploration of the historical relations between curriculum and pedagogy and the role of teachers in early childhood settings. We then turn to the ever expanding research base on early childhood teachers and teaching, illustrating how each strand of inquiry leads to a more complicated system of relations that form early childhood pedagogy. This chapter concludes with a discussion of the shifts that are taking place in the study of pedagogical sciences identifying directions for research and practice.

The Origins of Pedagogy for Young Children

Pedagogy is typically defined as the art and science of teaching. The term dates back hundreds of years to the Greek word "pedagogue." Originally, a pedagogue was a servant (often a slave) who attended to a young boy's ancillary educational needs such as carrying books and accompanying him to school (Monroe, 1913). The word is now synonymous with teaching and the techniques educators use to facilitate learning (Siraj-Blatchford, 1999). As Dewey (1938) noted, pedagogy is the medium that interconnects the child with the curriculum.

Research on teaching in the K-12 sector highlights how pedagogy is not a linear process from teacher to student but a relation that occurs in a context and that is composed of many elements. Watkins and Mortimer (1999) argue that pedagogy consists of relations between the teacher, the learner, the classroom, school and community contexts, content, and a theory of learning. More recently, philosophers and curriculum theorists (Farquhar & White,

2014; Hyun, 2006) in early childhood argue that pedagogy also includes moral, ethical, and emotional dimensions of teachers' work. For these latter scholars, pedagogy cannot be simplified to actions taken in a classroom but is also shaped by discourses about what it means to teach and learn, and these discourses shape power relations and teacher and learner subjectivities within any pedagogical encounter (Farquhar & White, 2014). In other words as the field of early childhood education has drawn on a broader range of theories to conceptualize early childhood education, definitions of pedagogy have expanded.

Despite the plethora of definitions of pedagogy by some scholars, early childhood teaching has often been distilled to child-centered or progressive pedagogy (Siraj-Blatchford, 1999; Farquhar & White, 2014). The notion of a child-centered education has its roots in the earliest philosophies and curriculum architects of early childhood education, including Froebel and his view of the kindergarten as a child's garden, and Rousseau's romantic notions of children developing at their own pace.

Over the twentieth century images of children who are in charge of their own learning have been supported and extended with various psychological theories of how young children develop. The most influential of these psychologists has been Piaget, whose notions of the child actively constructing his/her knowledge through interaction with both physical and social worlds has become the signature of what early childhood education is about. His stages of cognitive development have helped to contribute to the field being defined as birth through age 8 years or the ages that correspond with Piaget's sensory motor, preoperational, and concrete operational stages.

Building on Piaget's work has been the adoption of many of the principles of Vygotsky, who has argued that children learn best when working through the zone of proximal development or the space between what a child can do independently and what s/he can do with the help of others. Language is the tool through which children internalize their learning with others and play is considered to create its own zone of proximal development. Concurring with Piaget, Vygotsky argues that children build their knowledge of complex concepts as they interact with others within a socio-cultural context. However, from a Vygotskian perspective the teacher is less a facilitator than a leader of learning; yet, because both psychologists argue that the child is an active constructor of knowledge, it would seem that the image of the child having autonomy and input into his/her learning has been fundamental to early childhood educators' views of pedagogy.

According to Siraj-Blatchford (1999), because early childhood educators have embraced the notion of child-centered education, they often do not view pedagogy as a part of their role, instead emphasizing that they support children's learning; they do not teach. And this image of the teacher as a supporter or facilitator of children's learning has been codified in documents such as the "Guidelines for Developmentally Appropriate Practice" (Copple & Bredekamp, 2009), which position the child-centered educator in opposition to developmentally inappropriate practices where teachers are more explicit about content and what children can say and do throughout the day. To be sure there are variations on this dichotomy. The notion of intentional teaching Epstein (2014) explores how teachers might teach content in ways that support young children's learning. The starting point for every section of the book, however, is how teachers must base their pedagogy by starting with the child and knowledge of their development. Perhaps it is not surprising then that the scientific study of early childhood pedagogy, which places teachers and teaching as the central focus of research, has only become a line of inquiry in the field in the past two decades (Genishi, Ryan, Ochsner, & Yarnall, 2001). That is, until recently teaching has not been a major focus of research in the education of children from birth through age 8 years.

Research on Early Childhood Pedagogy

Despite teachers always being present in early childhood settings, what they do has often been subsumed within studies of curricula or specific kinds of early childhood programming. In other words, teaching has been reduced to specific characteristics of teachers such as qualifications, years of experience, etc. or characteristics of organizations such as teacher turnover and then these characteristics have been examined in relation to program quality or the impacts of particular curricula. Drawing on the work of researchers interested in teaching in K-12 settings, early childhood pedagogy is increasingly becoming a field of inquiry. As with evolving definitions of pedagogy, scientific investigations into early childhood teaching have included studies of pedagogy and cognition, studies of pedagogy as teaching through interactions, and studies of pedagogy as discursive and embodied. In what follows, we examine each of these differing programs of research on early childhood pedagogy, identifying methodological traditions and what has been learned about teaching young children.

Pedagogy and Cognition

In the 1980s scholars interested in teaching began to focus on teachers' thinking about various pedagogical issues and their reflections on practice. The earliest studies of teacher cognition in early childhood education tended to focus on studies of teachers' beliefs about developmentally appropriate practices; in part because the first published guidelines for developmentally appropriate practice (Bredekamp, 1987) were released in 1987 in an effort to challenge perceived academic pushdown. Several tools were developed and validated by Charlesworth and colleagues (Charlesworth, Hart, Burts, Thomasson, Mosley, & Fleege, 1993) and others (e.g., Snider & Fu, 1990) to document whether early childhood educators were implementing more child-centered practices over didactic instruction. Many of these studies examined the relationship between teacher characteristics and their self-reports of implementing developmentally appropriate practice, considered by the field as the signature of high-quality early childhood programming. For example, Snider and Fu (1990) used vignettes of classroom practice to elicit the beliefs of 74 child care teachers about best practice with young children. They found that teachers' education in child development and specialized early childhood coursework influenced their beliefs about developmentally appropriate practice.

Since the 1990s there have been a number of studies focused on teacher self-reports about their pedagogy. The topics are extensive and include studies of teachers' beliefs about dual language learners (Piker & Kimmel, 2018), the teaching of subject matter such as science (e.g., Pendergast, Lieberman-Betz, & Vail, 2017), and literacy (e.g., McKenney & Bradley, 2016) as well as studies of early childhood teacher beliefs about child-centered education (e.g., Cheung, Ling, & Leung, 2017; Graue, Whyte, & Delaney, 2014; Perren, Hermann, Illjuschin, Frei, Korner, & Sticca, 2017). A number of teacher belief studies are international comparisons focusing on how cultural context mediates teachers' cognitive commitments to things like risky play (e.g., Little, Sandseter, & Wyver, 2012) or the teaching of music (Hae, 2013), among other topics. Teacher belief studies have been conducted with both preservice and inservice teachers and some even compare novice teacher beliefs about a subject with those of experienced educators, with the aim of understanding the development of teacher beliefs over time.

However, self-reports via questionnaire or survey, or interviews alone, do not necessarily mean teachers do what they say in practice. Therefore some research on teacher cognition has shifted into investigations of teachers' beliefs and practices. Some of this research focuses on whether what teachers say actually translates into action in the early childhood classroom. For example, Wen, Elicker, and McMullen (2011) used a survey to elicit teacher self-reports about their curriculum practices and then observed these teachers in action employing the Early Childhood Teacher Behavior Assessment. They found that teacher beliefs and practices were weakly correlated. However, they did find that teachers with more years of experience and professional learning who believed there was a place for more teacher-directed strategies also enacted these beliefs in practice.

More recently, researchers have concentrated less on whether there is congruence between teacher beliefs and actions and more on the factors that mediate this relationship. For example, with the implementation of public preschool across the United States, researchers drawing on qualitative traditions have examined why teachers who may hold developmentally appropriate beliefs change their practices to fit in with K-12 policy mandates. In a qualitative case study of the enactment of preschool policy from children through to state policy actors in two state contexts, Graue, Ryan, Nocera, Northey, and Wilinski (2016) found that, despite each state having early learning standards based on developmentally appropriate practice, the majority of preK teachers felt they had no choice but to align at least part of their curriculum and teaching with K-12 standards by incorporating more direct instruction in academic content.

The broad range of studies on teacher cognition illustrates that what teachers think and do is a complex set of relationships. While teachers may hold implicit theories about pedagogy, their teaching interactions are shaped by a number of factors including policy, education, experience, and programmatic context.

Pedagogy as Teaching through Interactions

Researchers have recently conceptualized an interactional form of pedagogy as classroom quality, or "teaching through interactions" (TTI) (Hamre, Pianta, Downer, et al., 2013). In contrast to structural notions of quality that focus on regulatable factors like class size or teacher education that are inputs to education (Stuhlman & Pianta, 2009) or student outcome-focused measures that are outputs (Hanushek, 2002 in Hamre et al., 2013), the TTI approach frames quality as process, a pedagogy that is based on the essential interchange between the teacher and students. Intentionally not framing quality as a characteristic of the teacher, TTI take quality/pedagogy as an ecology that creates a learning context. From this perspective, child learning is most clearly understood through a framework describing the ways that social and emotional functioning are mediated through classroom interactions (emotional support), how classroom practices support autonomous management of student actions and paying attention (classroom organization), and strategies that support the construction of knowledge that is richly interconnected, contingent, and organized (instructional support). This model, enacted in the Classroom Assessment Scoring System (CLASS; Pianta, La Paro, & Hamre, 2008), does not take a content-specific approach to pedagogy and has only small differences across developmental levels from infancy to high school, illustrating the belief that the core teacher–student interactions have common elements regardless of student age. Research on the TTI framework has focused on the ways that all three domains are related to student outcomes ranging from behavior to academic performance.

Conceptualizing pedagogy as teaching through interactions is an important step in building a science of pedagogy in early childhood education. Describing a responsive pedagogy, it emphasizes teaching strategies that are contingent on specific students. It is premised on developmental frameworks recognizing that teaching is more than behaviors or curriculum.

But its reliance on those developmental frames is also a potential weakness for critical scholars. Conceptualizing developmental thought and theory as normatively based on white, middle-class values and cultural practices, critics worry that the pedagogy it promotes eliminates attention to the local, the cultural, the historical, and the linguistic (Delaney, 2018). By setting a single standard for how we teach, even if it is based on interactions, it standardizes a very human act. This is a particular concern if the pedagogy is ill fitted to minoritized students with different cultural practices. These concerns have prompted development of pedagogical models that expressly examine teachers' interactions with racially minoritized students, who have historically experienced oppression and teacher bias that limits their opportunities. Rather than assuming that pedagogy is culture free or that the instructional and the cultural are separate considerations, high quality teaching for minoritized students must consider both (Tharp, 1989). This makes sense given that mainstream instructional practices do just that for mainstream students. The Classroom Assessment of Sociocultural Interactions (CASI) and the Assessing Classroom Sociocultural Equity Scale (ACSES) were both designed to assess teacher use of instructional strategies or pedagogy that support racially minoritized students. ACSES focuses on pedagogies of equity and cultural alignment while CASI is also centered on alignment but also student agency and communal identity. Reflecting the integrated view described above, these models of pedagogy are seen as potentially supplemental to the TTI framework of CLASS. Coming from the same tradition as CLASS, these tools use observation instruments to drive pedagogy for particular purposes, using measurement based improvement to enhance teaching and interaction.

Pedagogy as Embodied and Discursive

Over the past two decades a scholarly movement has challenged the developmental knowledge base underpinning teaching practices in early childhood education. Employing a range of critical theories (Blaise & Ryan, 2020), a group of reconceptualist scholars have argued that developmentalism is not a theory of pedagogy (Silin, 1995) and that the field must examine teaching practices as discursive and political. This line of inquiry has led to a number of qualitative studies that look at the discourses teachers speak and put into action.

A large amount of this critical research on teaching examines the ways gender discourses are taken up and spoken into action. For example Blaise (2005) employed feminist post-structuralism to trace the gendered narratives children were enacting and how their feminist teacher challenged dominant understandings of what it means to be a boy and a girl. More recently, D'Souza Juma (2017) conducted participatory action research with 12 early childhood teachers in Pakistan showing how religious patriarchy shaped the work of these teachers. Using feminist poststructural concepts, these teachers were able to identify how religious discourses constrained their efforts to enact gender equity work with children. Aside from gender, critical scholars have used queer theory to consider teaching and sexuality (e.g., Blaise & Taylor, 2012; Silin, 1997) and postcolonial theory to investigate race and ethnicity (Gupta, 2006: Viruru, 2001). Several scholars have also used critical theories to investigate how policies shape teacher subjectivity and their ability to act as decision makers (e.g., Osgood, 2012).

A newer line of critical work has begun to conceptualize pedagogy as embodied. According to Tobin and Hayashi (2015), this line of inquiry comes from post humanist scholars questioning the importance placed on mind and language over the material. From this perspective, teaching involves connecting the mind and body, emotion, and intellect. Embodied pedagogy involves paying attention to both one's own body and those of others, and the meanings physical movement like gesture, facial expressions etc. convey to young children. Moreover, aspects of the material world, such as objects, physical space etc., are just as important as the human world. According to Tobin and Hayashi (2015), embodied pedagogy "visualizes teaching as emerging out of the interaction of minds, bodies, things, ideas, and contexts" (p. 327).

To get at the techniques of the body used by educators, some scholars are using video with teachers. For example Tobin and Hayashi (2015) were interested in the cultural practices of Japanese teachers. They showed the teachers video of themselves, first asking them about specific interactions and then examining each scene for bodily techniques such as posture, gaze, and touch. In doing so, they were able to see with the teachers in their study how particular bodily actions had effects on children's play. In another qualitative study using video (Xiao & Tobin, 2018), student teachers were required to video themselves for assessment purposes and were encouraged to focus on embodied aspects of their teaching as they watched themselves. While Xiao and Tobin found that this approach was effective in helping preservice teachers to think about how they used their bodies when teaching, they also report that for some of the preservice teachers having to video themselves made them anxious, resulting in their choosing to teach from the front of the classroom in more traditional ways.

Research on pedagogy as discursive and embodied provides lenses to help teachers to consider how they position themselves in time and space and in relation to objects as well as children in all parts of teaching. Moreover, these studies show that teaching is not always benign but that pedagogy is also political, shaping children's and teachers' identities. This work has introduced new theoretical concepts to the study of pedagogy, illuminating the complexity of teaching young children. However, while this work is cutting edge, it remains to be seen how these theories will be taken up and employed by teachers.

Concluding Thoughts

It has been argued for over a century whether pedagogy is a science. For scholars who have taken pedagogy as an empirical question, that argument has been irrelevant and they have pursued multiple approaches to understanding what pedagogy is and how it comes to be. For some, the question is about whether early childhood pedagogy is a cognitive process that can be studied through teacher belief (e.g., Cheung et al., 2017). Or it has been disassembled into components that can be described, rated in practice, and used for professional education (e.g., Hamre et al., 2013). Others argue that pedagogy is contingent on the social, cultural, and political contexts (e.g., Graue et al., 2016) and that it is not possible to generalize pedagogical principles across contexts and cultures.

However, as we argue in this chapter, the science of early childhood pedagogy is very much a robust field of study with an exciting and productive range of programs of inquiry. These strands of research are leading the field in new directions by placing educators and teaching at the forefront of investigations about the education of young children. The key, we think, to building a generative science of pedagogy is to respect the knowledge potential of these diverse programs of inquiry and to engage across them. We have a way to go in that regard but we hope that this book is a step in the right direction.

References

Blaise, M. (2005). *Playing it straight! Uncovering gender discourses in the early childhood classroom*. New York: Routledge.

Blaise, M., & Ryan, S. (2020). Engaging with critical theories and the early childhood curriculum. In J. J. Mueller & N. File (Eds.), *Curriculum in early childhood education: Re-examined, reclaimed, renewed* (2nd ed.) (pp. 80–95). New York: Routledge.

Blaise, M., & Taylor, A. (2012). Using queer theory to rethink gender equity in early childhood education. *Young Children*, 67(1), 88–96.

Bredekamp, S. (1987). *Developmentally appropriate practice in early childhood programs serving children from birth through age 8*. Washington, DC: NAEYC.

Charlesworth, R., Hart, C.H., Burts, D., Thomasson, R., Mosley, J., & Fleege, P. (1993). Measuring the developmental appropriateness of kindergarten teachers' beliefs and practices. *Early Childhood Research Quarterly*, 8(3), 255–276.

Cheung, S.K., Ling, E.K., & Leung, S.K.Y. (2017). Beliefs associated with support for child-centered learning environments among Hong Kong preservice early childhood teachers. *Journal of Education for Teaching*, 43(2), 232–244.

Copple, C., & Bredekamp, S. (2009). *Developmentally appropriate practice in early childhood programs serving children from birth through age 8*, 3rd edn. Washington, DC: NAEYC.

Delaney, K.K. (2018). Looking away: An analysis of early childhood teaching and learning experiences framed through a quality metric. *Contemporary Issues in Early Childhood*, 19(2), 167–186. doi:10.1177/1463949118778023

Dewey, J. (1938). *Experience and education*. Kappa Delta Pi Lecture Series. New York: Collier/MacMillan.

D'Souza Juma, A. (2017). Engaging with feminist poststructuralism to inform gender equity practice in early childhood classrooms in Pakistan. In K. Smith, K. Alexander, & S. Campbell (Eds.), *Feminism(s) in early childhood: Perspectives on children and young people*, vol. 4. Singapore: Springer.

Epstein, A. (2014). *The intentional teacher: Choosing the best strategies for young children's learning* (revised ed.). Washington, DC: National Association for the Education of Young Children.

Farquhar, S., & White, E.J. (2014). Philosophy and pedagogy of early childhood. *Educational Philosophy and Theory*, 46(8), 8221–8832, doi:10.1080/00131857.2013.783964

Genishi, C., Ryan, S., Ochsner, M., & Yarnall, M. (2001). Teaching in early childhood education: Understanding practices through research and theory. In V. Richardson (Ed.), *Handbook of research on teaching* (4th ed., pp. 1175–1210). Washington DC: American Educational Research Association.

Graue, E., Whyte, K., & Delaney, K.K. (2014). Fostering culturally and developmentally responsive teaching through improvisational practice. *Journal of Early Childhood Teacher Education*, 35(4), 37–41. https://doi.org/10.1080/10901027.2014.968296

Graue, M.E., Ryan, S., Nocera, A., Northey, K., & Wilinski, B. (2016). Pulling pre-K into a K-12 orbit: The evolution of pre-K in the age of standards. *Early Years*, 37, 108–122.

Gupta, A. (2006). *Early childhood education, postcolonial theory, and teaching practices in India: Balancing Vygotsky and the Veda*. New York: Palgrave Macmillan.

Hamre, B.K., Pianta, R.C., Downer, J.T., DeCoster, J., Mashburn, A.J., Jones, S.M., … Hamagami, A. (2013). Teaching through interactions. *The Elementary School Journal*, 113(4), 461–487. https://doi.org/10.1086/669616

Hae, K.K. (2013). A comparison of early childhood preservice teachers' beliefs about music and developmentally appropriate practice between South Korea and the US. *Australasian Journal of Early Childhood*, 38(2), 122–128. doi:10.1177%2F183693911303800215

Hyun, E. (2006). Transforming instruction into pedagogy through curriculum negotiation. *Journal of Curriculum and Pedagogy*, 3(1), 136–164, doi:10.1080/15505170.2006.10411587

Little, H., Sandseter, E.B.H., & Wyver, S. (2012). Early childhood teachers' beliefs about risky play in Australia and Norway. *Contemporary Issues in Early Childhood*, 13(4), 300–316.

McKenney, S., & Bradley, B. (2016). Assessing teacher beliefs about early literacy curriculum implementation. *Early Childhood Development and Care*, 186(9), 1415–1428.

Monroe, P. (1913). *A cyclopedia of education*. New York: Macmillan.

Osgood, J. (2012). *Narratives from the nursery: Negotiating professional identities in early childhood.* New York: Routledge.

Pendergast, E., Lieberman-Betz, R., & Vail, C. (2017). Attitudes and beliefs of preK teachers toward teaching science to young children. *Early Childhood Education Journal*, 45(1), 43–52.

Perren, S., Hermann, S., Illjuschin, I., Frei, D., Korner, C., & Sticca, F. (2017). Child-centered educational practice in different education settings: Associations with professionals' attitudes, self-efficacy and professional background. *Early Childhood Research Quarterly*, 38, 137–148.

Pianta, R.C., La Paro, K.M., & Hamre, B.K. (2008). *Classroom assessment scoring system (CLASS).* Baltimore: Brookes.

Piker, R., & Kimmel, M. (2018). Preparing young dual language learners for school success: early childhood teacher beliefs regarding school readiness. *Early Childhood Development and Care*, 188(10), 1368–1380.

Silin, J. (1995). *Sex, death and the education of our children: Our passion for ignorance in the age of AIDS.* New York: Teachers College Press.

Silin, J. (1997). The pervert in the classroom. In J. Tobin (Ed.), *Making a place for pleasure in the early childhood classroom* (pp. 214–234). New Haven: Yale University Press.

Siraj-Blatchford, I. (1999). Early childhood pedagogy: Practice, principles and research. In P. Mortimer (Ed.), *Pedagogy and its impact on learning* (pp. 20–45). London: Paul Chapman Publishing.

Snider, M., & Fu, V. (1990). The effects of specialized education and job experience on early childhood teachers' knowledge of Developmentally Appropriate Practices. *Early Childhood Research Quarterly*, 5 (1), 69–78.

Stephen, C. (2010). Pedagogy: The silent partner in early years learning. *Early Years: An International Research Journal*, 30, doi:10.1080/09575140903402881

Stuhlman, M.W., & Pianta, R.C. (2009). Profiles of educational quality in first grade. *The Elementary School Journal*, 109(4), 323–342. doi:10.1086/593936

Tharp, R.G. (1989). Psychocultural variables and constants: Effects on teaching and learning in schools. *American Psychologist*, 44(2), 349–359. doi:10.1037/0003-066X.44.2.349

Tobin, J., & Hayashi, A. (2015). Using video for microanalysis of teachers' embodied pedagogical practices. *Research in Comparative and International Education*, 10(3), 326–336.

Viruru, R. (2001). *Early childhood education: Postcolonial perspectives from India.* New Delhi, India: Sage.

Watkins, C., & Mortimer, P. (1999). Pedagogy: What do we know?In P. Mortimer (Ed.). *Pedagogy and its impact on learning* (pp. 1–19). London: Paul Chapman Publishing.

Weber, E., Hart, C., Burts, D., Thomasson, R., Mosley, J., & Fleege, P. (1993). Measuring the developmental appropriateness of kindergarten teachers' beliefs and practices. *Early Childhood Research Quarterly*, 8(3), 255–276.

Wen, X., Elicker, J., & McMullen, M. (2011). Early childhood teacher beliefs: Are they consistent with observed classroom practices? *Early Education and Development*, 22(6), 945–969.

Xiao, B., & Tobin, J. (2018). The use of video as a tool for reflection with preservice teachers. *Journal of Early Childhood Teacher Education*, 39(4), 328–345.

11

THE CRITICAL CONTRIBUTIONS OF ASSESSMENT SCIENCE TO EARLY CHILDHOOD EDUCATION

Kathleen Hebbeler and Megan E. Cox

The valid measurement of what young children know and can do is essential to advancing developmental science, as well as knowledge of effective intervention, instruction, and programs. Without assessments, there is no science of early childhood education. Assessment is the process of gathering information for the purpose of making decisions about an individual, group of individuals, or program (American Educational Research Association, American Psychological Association, & National Council on Measurement in Education, 2014; Bagnato, 2007; Meisels, 1996; National Research Council, 2008). In early childhood education, as in other social and educational sciences, we use measurement to assign a score to a child's characteristic such as a skill or a trait.

Assessment in early childhood education can measure many things, be conducted in markedly different ways, and for very different purposes. Different sources characterize the various purposes of early childhood assessment in slightly different ways, but common themes emerge. Purposes for early childhood assessment include: screening and diagnosis for the identification of special needs; individual program planning and monitoring to guide intervention and to support learning; social benchmarking or monitoring of trends which portrays the overall status of young children; accountability which provides information about public investments; and program evaluation and research which seeks to determine effectiveness of early childhood programs and advance knowledge (Kagan, Rosenkotter, & Cohen, 1997; McLean, Wolery, & Bailey, 2004; Nagle, 2007; National Research Council, 2008; Shephard, Kagan, & Wurtz, 1998). The first two purposes, identification of special needs and supporting learning, involve making decisions about individual children and thus involve the use of information about a single child whereas the other purposes refer to decisions about groups of children and use aggregated data.

A great deal has been written about the use of assessment for supporting teachers and other practitioners in individual decision making. In this chapter, we focus on the assessment of young children for the purposes of conducting early education program evaluation and research. The role of assessment in furthering our understanding of early childhood education has received far less attention and, in fact, is even omitted from some listings of the multiple purposes of assessment with young children. Space limitations will not allow even an introductory discussion of the science of assessment development. An extensive literature exists on this subject and there is nothing unique in the application of science to the development of assessment tools for young children. There are, however, many issues that are unique to the assessment of young children and resulting challenges of how to interpret the findings of those assessments in the context of advancing the science of early childhood education. We will use the phrase "early childhood education" to refer to the full range of early care and education programs, including prekindergarten programs supported by public or private funds, Head Start, and center-based child care programs, because all of these programs focus to some degree on promoting children's early learning (Kagan & Kauerz, 2007).

Program Goals, Constructs, and Assessments

The power of assessment to advance knowledge about early childhood education rests on the connections between program goals, outcome constructs, and assessment measures. Figure 11.1 presents a generic and common study design that examines the impact of a program or program feature on a child outcome. Studies seek to examine a relationship between a program (e.g., State PreK) and an outcome construct (e.g., school readiness), which is then measured by an assessment. A more targeted study might explore one or more features of an early childhood program (e.g., the curriculum, the length of the day) and one or more child outcomes. Other variations may include asking about multiple program features, multiple outcomes, examining long- as well as short-term outcomes or the impact of mediating and moderating variables. The underlying relationships depicted in the graphic remain the same for these more complex questions.

The child outcome construct boxes contain the countless types of knowledge, abilities, skills, behaviors, attitudes, and dispositions that early childhood programs could impact. These outcomes include a hierarchy of superordinate (e.g., language) and multiple layers of subordinate constructs (vocabulary, receptive and expressive vocabulary, academic vocabulary, etc.) as we will discuss in the next section. The program's goals should be reflected in the outcome constructs identified as the focus of program effectiveness studies. In some studies, the outcome construct is obvious; in a program designed to improve self-regulation, the construct of interest is self-regulation. In many large-scale evaluations in early childhood, the outcomes are considerably more generic and less well specified. For example, many early childhood programs are designed to improve school readiness, which is a global construct whose subordinate constructs will need to be specified for measurement purposes.

The realm of child assessment is captured in the lower right boxes but the foundation for any study is the conceptual model in the top half. Specification of the outcome constructs will or at least should drive the assessment choices. A simple but key point is that the assessment is not the same as outcome construct. Assessment is a technique for collecting data on the construct. Just as the outcome constructs must be aligned with the program's goals, the assessments utilized must be well aligned with and encompass the full breadth of the intended outcomes because what is measured by the assessments become the de facto outcome constructs. Every study looking at child outcomes is totally dependent on what is being assessed and how well the assessments perform.

To illustrate the implications of the construct–assessment relationship, let's assume that Program Feature A improves Child Outcomes X, Y, and Z. Any study's ability to discover this relationship depends on the study design correctly identifying X, Y, and Z as the target constructs at both the superordinate and subordinate outcome level and assessing them well. If a program seeks to improve literacy, social skills, and executive functioning but only literacy is measured or well measured, some impacts will be missed. Even worse, if the program impacts social skills and executive functioning but not literacy, the erroneous conclusion would be that the program has no impact. Within literacy, if the study identifies alphabet knowledge as the key construct but the "real outcome" is vocabulary, again the true impact of the program remains undiscovered.

When a study fails to find a relationship, a misstep in any one of the boxes in the graphic could be responsible. Assessment tools that are slightly misaligned with the construct, that measure only one sub-area, or that present other validity challenges, seriously limit a study's ability to detect a true relationship between program and outcome. As we will discuss, selection of constructs and assessments in early childhood research presents many challenges, including that some assessments poorly reflect developmental science and traditional assessment methods are ill suited to the nature of the young child. Given the multiple opportunities for real impacts to be missed, it is quite remarkable that a substantial body of evidence supporting the effectiveness of early childhood education has been established.

What to Measure

Our understanding of the constructs that should be measured to determine the effectiveness of early childhood programs has evolved significantly over the last 50 years. This evolution has been driven by, as well as contributed to, advances in knowledge about young children's learning and development.

An Overemphasis on Cognition

Early childhood education as public policy emerged in the 1960s in response to the troubling issue that children from low-income families showed poorer academic performance in school than their more advantaged peers (Bishop-Josef & Zigler, 2011). Based in the emerging scientific knowledge of the impact of early environment on later development, Head Start was implemented as a federal policy solution to this problem. From the beginning, Head Start was a comprehensive program providing screenings, mental health services, nutrition education and other services in addition to an educational program (Zigler, Styfco, & Gilman, 1993; Zigler & Styfco, 2000). As a major federal investment, Head Start would become the subject of many evaluations.

At roughly the same time, several groups of university researchers were examining whether well-designed early learning environments could improve development for preschoolers from low-income families by implementing smaller and often more intensive and controlled early childhood programs (Lazar et al., 1982). These studies included the now well-known Carolina Abecederian and the Perry Preschool Projects (Ramey & Campbell, 1991; Schweinhart & Weikart, 1981). Drawing on the same scientific evidence as Head Start about the role of the child's environment on development, these studies sought to learn if they could engineer an enriched environment that would positively impact development for children who were likely to show poor outcomes. In a more orderly world, the experimental studies would have provided the basis for Head Start but as it turned out they were both exploring the same question at the same time: could early education improve child outcomes for children in poverty?

The initial program goals for Head Start were quite vague although an early planning document laid out several objectives for the program. These included improving the child's mental processes and skills, and encouraging self-confidence, spontaneity, curiosity, and self-discipline (Zigler & Styfco, 2000). Because Head Start lacked clearly defined program outcomes in its initial years, the outcomes were defined by what the program evaluations measured. IQ tests became the principal measure of the success for Head Start and for the university studies. At the time, IQ was a common outcome measure in developmental research. IQ tests were well-developed, had good psychometric properties, were easy to administer, and were known to predict school performance (Zigler & Trickett, 1978). Ironically, IQ was considered a stable trait, which made it an odd choice as an outcome measure for a program trying to change outcomes.

The exclusive focus on the intellectual component of Head Start is an early example of the misalignment of program goals and outcomes measures. IQ was never an accurate reflection of Head Start's whole child approach. This misalignment along with findings from a major study that showed the early advantages for Head Start graduates were not sustained into early elementary led to early questions about IQ as an appropriate outcome measure. In subsequent years, IQ tests have been widely criticized as a of measure program effects because they are not useful measures of growth and may be biased against racial and cultural minority groups (Meisels & Atkins-Burnett, 2000; Ramey & Ramey, 2004; Shonkoff & Phillips, 2000) Also, score changes have been found to be influenced by factors like motivation, which have nothing to do with intelligence (Zigler & Styfco, 2000).

The demise of IQ raised the question of what should be used instead. Consistent with a recommendation by Zigler and Tricket (1978) that social competence should be seen as the intended outcome, the federal office responsible for overseeing Head Start adopted this construct as the outcome for future evaluations. Social competence is loosely defined as the qualities and abilities children need to succeed in an academic environment. The agency invested heavily in operationalizing the construct and developing measures for use in program evaluations but encountered several barriers and, after ten years of work, the funding ended in the early 1980s without the development of a tool to measure it (Schrag, Styfco, & Zigler, 2004).

Multiple Dimensions

A new way of conceptualizing child outcomes emerged in the 1990s that would exert a substantial influence on research and program evaluations as well as policy and practice that continues to the current day. The National Education Goals Panel (NEGP) was created in 1990 to measure progress toward six national goals that were part of plan to address low

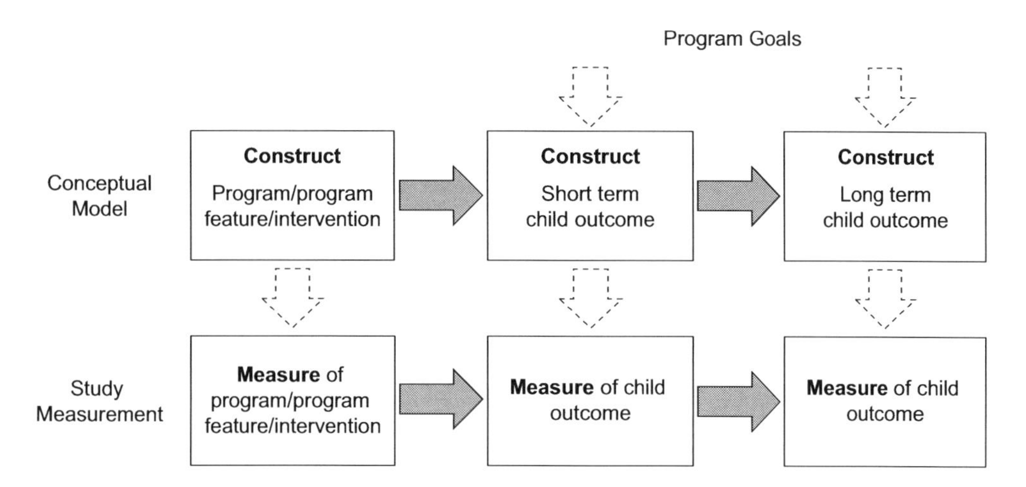

FIGURE 11.1 Relationships between Program Goals, Constructs, and Measures

achievement in America's schools. The first goal was an early childhood goal: "By the year 2000, all children in America will start school ready to learn" (National Education Goals Panel, 1991b).

A technical planning group was formed to design a process for measuring progress toward Goal 1, referred to as the readiness goal. The work group was critical of past attempts to assess "preparedness for school" with a one-dimensional emphasis on general knowledge and abilities. Their conception of learning and development, in accord with the prevailing thinking in developmental science, was multi-faceted, describing the whole child rather than a single element. The planning group use the word "dimensions" to describe five construct areas proposed for measurement: (1) physical well-being and motor development; (2) social and emotional development; (3) approaches toward learning; (4) language usage; and (5) cognition and general knowledge. The work group emphasized the unique and important contribution of each dimension, noting that they "cannot be aggregated in any way to form a single quotient" (National Education Goals Panel, 1991a, p. 8). A detailed plan for the development of a new assessment to measure the nation's progress on Goal 1 was developed but no funding was ever provided.

Despite the lack of progress toward an assessment strategy, the NEGP left a significant legacy for early childhood assessment. Controversies swirled around the meaning of school readiness (see, e.g., Graue, 1992; Schorr, 1997) but the concept became solidly embedded in the national conversation as a widely accepted outcome of early childhood education. Nevertheless, a firm conceptualization of school readiness remains elusive. At a global level, school readiness refers to skills and knowledge associated with success in school but the construct cannot be measured at this level (Blair et al., 2007; Snow, 2007). The NEGP's view of readiness as made up of multiple dimensions was accepted and has influenced measurement choices in early education research for decades. The NEGP's dimensions also provided a framework for states as they developed early learning guidelines for 3- to 5-year-olds and later for birth to 3-year-olds and thus in theory identified not only what should be taught but also what should be measured in state evaluations of their PreK programs (Scott-Little, Kagan, & Frelow, 2006; Scott-Little, Kagan, Frelow, & Reid, 2009).

The inclusion of social-emotional as one of the dimensions provided an appropriate counterbalance to the previous overemphasis on cognitive skills as measured by IQ and built on the groundwork that had already been laid with the earlier recognition of social competence. The call to measure social-emotional skills as an outcome of early childhood education has been reiterated many times over the years since the NEGP (Atkins-Burnett, 2007; Halle & Darling-Churchill, 2016; Hirsh-Pasek, Kochanoff, Newcombe, & de Villiers, 2005; Raver, 2002). Measuring social-emotional development, however, continues to be challenging because of a lack of consensus on the key subordinate constructs as well as the inherently contextual and cultural nature of social behavior (Campbell et al., 2016; Darling-Churchill & Lippman, 2016; Jones, Zaslow, Darling-Churchill, & Halle, 2016). For example, with regard to defining the construct, Halle and Darling-Churchill (2016) identified four sub-constructs (social competence, emotional competence, behavior problems, and self-regulation) whereas Hernandez et al. (2016) identified two (emotional development and social interaction). A small sample of the sub-constructs associated with social-emotional development include: emotional regulation, self-awareness, self-confidence, self-direction, cooperation, perspective taking, expressing emotion, emotional knowledge, and emotional expressiveness. Given the consensus of the importance of measuring social-emotional development when looking at program effectiveness, a clear need exists for more conceptual clarity that is a prerequisite to the identification or creation of good measures that can feasibly be used in research and evaluation.

Approaches toward learning, a new outcome introduced by the NEGP, was the least understood and least researched of the dimensions (Kagan, Moore, & Bredekamp, 1995). Approaches to learning is an umbrella construct for a set of attitudes, habits, and learning styles that describe the manner in which children go about the process of learning. It includes constructs such as curiosity, creativity, initiative, task persistence, imagination, and invention (Kagan et al., 1995). Like social-emotional, the lack of consensus on conceptualization has made measurement challenging and there are not good assessment tools for the array of sub-constructs under the approaches to learning umbrella (Love, 2001). Unlike social-emotional, approaches to learning has not found a groundswell of support for inclusion as an outcome in studies of early childhood programs. Some states, however, have included approaches to learning as an area within their early learning guidelines, signaling some ongoing recognition of its importance as an intended outcome of early education (Scott-Little et al., 2006).

New Developments

Advances in our understanding of learning and development have added new constructs as possible outcomes of early childhood education. One such construct is executive function, which has strong connections to research in neuroscience (Bierman, Nix, Greenberg, Blair, & Domitrovich, 2008; Phillips, Lipsey, et al., 2017; Raver & Blair, 2016). Executive function refers to a variety of interdependent skills that are necessary for purposeful activity (Shonkoff & Phillips, 2000). Inhibition including self-control, working memory, and cognitive flexibility are generally considered the core executive functions (Diamond, 2013). Although a relatively recent construct, much work has occurred around assessment. Ackerman and Friedman-Kraus (2017) identified 83 measures of executive function in children including single task measures, performance task batteries, and teacher rating scales. Executive function is likely to be included as an outcome in more early education research in the future, although such a dizzying assortment of measures means researchers face "measurement mayhem" (Morrison & Grammer, 2016).

Another scientific advance that has pointed to new areas for measurement is the growing body of knowledge on the early foundations for academic subject areas, most notably literacy and mathematics. The precursors for learning to read are not just general cognitive skills but a set of specific skills such as knowing that words are made up of letters and print is connected to speech. Examples of early literacy constructs include knowledge of conventions of print, alphabetic knowledge, phonological awareness, syntactic awareness, letter-sound correspondence, phonological memory, understanding of narrative, and vocabulary and other language skills (Snow & Oh, 2011; Whitehurst & Lonigan, 1998). Similarly, the foundation for learning formal mathematics in school is built through children's early experiences with counting, concepts of more-or-less, and other kinds of informal mathematics that are part of young children's everyday lives. Early math constructs address both content (i.e., numbers and operations; algebra; geometry; measurement; and data analysis) and process constructs (problem solving, communication, reasoning, representations and connections) (Cross, Woods, & Schweingruber, 2009).

Not all emerging skills in these areas may be equally suitable as candidates for program outcomes, however. Literacy researchers make a distinction between constrained and unconstrained skills (Paris, 2005; Snow & Matthews, 2016). Constrained skills are finite, readily teachable, and have a ceiling in that all children learn them. Learning the names of the letters of the alphabet is an example. Unconstrained skills such as vocabulary and background knowledge are acquired more gradually through experience, are more related to a child's background, and are more difficult to teach and test than constrained skills. The implications for assessment of outcomes are significant. Because constrained skills are learned over a window of time in a child's life, they go from no variance in the population (floor), to variance in skill while children are learning them, and back to no variance (ceiling). Paris (2005) maintained that research showing relationships between constrained skills and later reading is misleading, capturing a proxy variable such as a home environment where some children had an opportunity to learn the skill. This same logic argues against using constrained skills as an outcome measure of early childhood programs. Since all children learn them, it is inevitable that the children in the comparison group will acquire the same skills as the children in the experimental group. The assessment is only capturing who gets there first, which is of questionable value as an outcome. Constrained skills are important and need to be learned but unconstrained skills have more long-term meaning (Snow & Matthews, 2016). The concept of constrained skills is based in literacy research but applies equally well to early mathematics; knowing number words to 20 and counting 10 objects correctly are constrained skills. Bailey et al. (2017) suggest that intervention program needs to measure "trifecta skills," i.e., skills that are malleable, fundamental, and unlikely to develop in the absence of intervention. They conclude that very few early childhood skills for children living in the "normative range of environmental conditions" meet these criteria. Children living outside the normative range are those who have experienced trauma or other adverse environmental conditions. Intervention programs for these children seek to restore functioning, not help them to acquire malleable skills earlier.

How We Measure

Children show their knowledge, skills, processes, abilities, or traits through their behaviors. Researchers use various kinds of assessment techniques to capture these behaviors and then make inferences about the child's status regarding a construct based on a score. Examples of such techniques are asking a child to complete a series of tasks or asking an adult who knows the child to answer a series of questions about the child's behaviors. We will review some of the assessment techniques used in early childhood research and evaluation, explore their strengths and weaknesses, and discuss the possible implications for gathering accurate information about children's behaviors. Techniques for gathering information about young children differ from one another along several dimensions, including: the extent to which information is being collected in a context that is contrived or natural to the child; whether the behaviors are naturally occurring or are being elicited and, if elicited, whether by a familiar adult or by a stranger; and the extent of the standardization of the process. Given the nature of the young child, the more unnatural the situation is to the child, the lower the likelihood the child will display typical behavior or skills. All assessment approaches are based on a sample of behavior which is assumed to be representative of what one would learn if far more information were to be collected about the child. The representativeness of the behavior sample is crucial to accurately capturing the child's status on the outcome construct, which in turn is crucial to the accuracy of study findings based on an assessment.

One commonly used technique for learning about children is direct assessment. As typically used in research and evaluation, a direct assessment involves an unknown adult working one-on-one with the child in an unfamiliar setting asking the child to complete a series of tasks. The interaction is highly scripted with strict rules governing what the assessor is and is not allowed to say and do. Together, these features make the interaction completely unlike anything most young children have ever encountered. The logic behind standardization of administration is that it allows the child's results to be compared with those of other children who receive the same directions. The assumption is that comparable conditions produce consistency of test results and comparable score interpretation, including comparison with a norming sample. Administering a direct assessment to a young child will standardizes the spoken words and the activity, but it does not standardize how the child experiences the interaction and, in fact, different children may experience it completely differently.

Much has been written about the problems with administering direct assessments to young children (Atkins-Burnett, 2007; Meisels, 2007; National Early Childhood Accountability Task Force, 2007; National Research Council, 2000, 2008; Neisworth & Bagnato, 2004; Urban & Swadener, 2016). An oft-repeated conclusion is that young children are notoriously poor test takers in part because context is extremely important for young children's learning and behavior (National Institute of Child Health and Human Development, 2002). Being with an unfamiliar adult in the assessment situation may be anxiety provoking. Some children may have experienced being asked to answer a series of disjointed questions by adults (e.g., "show me the red one") but, for others, it might be a completely new experience and some of these differences are related to the child's culture. In some cultures, this kind of direct questioning would be very unusual or even rude. Children may be trying to figure out what the assessor wants them to say, or not be able to process the language being used. Young children have limited self-regulation skills, which means they can be easily distracted by the novelty of the test objects and not pay attention to the question. Strict standardization can be especially problematic for children with disabilities or children who are dual language learners with limited English skills.

The likely consequence of the developmentally inappropriate demand characteristics of the direct assessment situation would be non-random measurement error in the results. The children for whom performing tasks on cue for an adult is an unusual activity are most likely to be children from non-mainstream cultures or those who have difficulties attending or processing language; these children may miss items on skills that are within their repertoire. Our increasingly diverse preschool population has led to calls for more sub-group analyses to be able to answer questions about differential effectiveness (Phillips, Johnson, Weiland, & Hutchison, 2017; Shonkoff & Phillips, 2000). Answering these important questions will require more research on how well direct assessment performs with different sub-populations. To date, there is little research on bias in early childhood assessments (National Research Council, 2008).

Given these issues, it is not surprising that direct assessments administered in preschool and kindergarten are poor predictors of later school achievement (La Paro & Pianta, 2000). Nevertheless, direct assessments have been the assessment tool of choice in research on early education programs for many years. One reason is that they display good psychometric properties on traditional metrics, in part because the psychometric analyses are based on comparing direct assessments with other direct assessments that elicit the same kinds of behaviors from young children (Atkins-Burnett, 2007). Another extremely important strength of direct assessments is that they allow information to be collected efficiently on a large number of children.

There are alternatives to direct assessments. The National Association for the Education of Young Children & National Association of Early Childhood Specialists in State Departments of Education (NAEYC & NAECS/SDE, 2003), the Division for Early Childhood of the Council for Exceptional Children (2007), and the National Association of School Psychologists (NASP, 2015) agree on the type of assessment that should be used with young children. These organizations have called for assessment based on a child's everyday routines, interest, and materials that capture naturally occurring skills and behavior. The NASP (2015) position statement urges school psychologists to collect data in natural settings. Methods that incorporate familiar context, familiar activities, and familiar adults include rating scales, observation-based assessments, and samples of the child's work. These methods capture children's behaviors and skills in everyday environments such as an early childhood classroom or home where the target knowledge, skills, and abilities occur naturally across a variety of activities (Bagnato, 2007; Bodrova & Leong, 2018; Meisels, 1996). Presumably, they also reflect a larger sample of behavior than what is assessed in the few tasks in a direct assessment. These methods involve some degree of standardization to achieve reliability but do not involve the scripted behavioral interactions of direct assessment. Although more suited to the nature of the young child, these methods present their own set of challenges for use in research and evaluation.

Rating scales involve an adult familiar with the child such as a teacher or parent responding to a series of questions or prompts about the child's behavior. Rating scales tend to be used for assessing social-emotional constructs, many of which do not lend themselves to direct assessment. A recurring finding in the literature is that parent and teacher ratings of the same child differ (Marcella & Howes, 2015). For example, Konold and Pianta (2007) studied child behavior ratings of mothers, fathers, and teachers and found results were strongly influenced by the informant. Explanations for parent–teacher differences include that one is the better rater or that children behave differently at home and in classrooms and that different ratings provided by parent and teacher reflect that child in different contexts. Suen, Logan, Neisworth, and Bagnato (1995) found that using convergence models to combine parent and teacher ratings produced a more reliable picture of the child's abilities and characteristics than either rating by itself. Pooled information also aligns with recommendations of the professional organizations cited above to use multiple informants to measure a child's skills and abilities. Parents and teachers may contribute different information but the evidence suggests both can be accurate and reliable raters when behavior descriptors are well-defined (Perry & Meisels, 1996; Suen et al., 1995). More research is needed to understand rating scale characteristics, context, rater differences, and influences on raters.

Observation-based measures assess children's knowledge, skills and behaviors within a familiar context and everyday activities. This type of assessment involves an observer "scoring" a set of indicators analogous to the items on a direct assessment. A good observation tool will provide a set of behavioral descriptors that the observer uses to determine the presence (e.g., yes–no) or extent of the behavior or skill (e.g., not yet emerging–present) based on watching the child (Bodrova & Leong, 2018). Observation-based assessments can be completed by an adult familiar to the child or by a researcher. Data can be either in real time or obtained through the coding of video. Some tools are designed to include information gathered from multiple sources such as parents or other teachers, which can give the assessor additional information to inform the scoring. Commercially available observation-based assessments are used by teachers to guide instruction but can be used in research. For example, an observational tool was used to measure children's school readiness in an evaluation of the effects of full vs part-day PreK on children's outcomes (Reynolds et al., 2014).

Some research on teachers' use of observation-based has concluded that this type of assessment does not produce valid information especially when compared with direct assessment (Waterman, McDermott, Fantuzzo, & Gadsden, 2012). For instance, Waterman and colleagues (2012) contend that teachers may find it difficult to maintain focus on relevant behaviors when observing children in the classroom or may have trouble with objectivity. Some of these studies and their conclusions are problematic because the amount of training the teachers have received on the use of the assessment is either minimal or it is not known, with no information present on whether inter-rater reliability was established. The direct assessments, on the other hand, were administered by research staff who were trained to a high standard as part of the research (Russo, Williford, Markowitz, Vitiello, & Bassok, 2019; Waterman et al., 2012).

In contrast, other research concludes that observation-based measures do produce valid and reliable information. (Atkins-Burnett, 2007; Cabell, Justice, Zucker, & Kilday, 2009) (Meisels, Bickel, Nicholson, Xue, & Atkins-Burnett, 2001; Perry & Meisels, 1996). Necessary conditions for tools to yield valid results are well-defined indicators and the assessors are well trained in how to collect the information (Downer, Booren, Lima, Luckner, & Pianta, 2010; Lambert, Kim, & Burts, 2014; Perry & Meisels, 1996; Sekino & Fantuzzo, 2005; Soderberg, Stull, Cummings, Nolan, & McCutchen, 2013). Additional research is needed to understand the training necessary to reach reliability on observational measures and the conditions under which valid inferences can be made for research and evaluation purposes.

Portfolios of children's work samples can provide insight into children's progress over time in a classroom (Darling & Grace, 2015). A portfolio is a collection of a child's work in a given area. Portfolios reflect both the process and product of a child's learning (Epstein, Schweinhart, & DeBruin-Parecki, 2004). Criteria need to be identified for a portfolio system, including establishing the purpose and criteria for collecting student work and the types of work to be collected (Shores & Grace, 1998). Although work samples are a rich and authentic source of information on what children can do, they are a resource intensive data collection strategy in a study because they involve training teachers on what to collect, collecting products over a period of time, and applying a detailed coding system to those products.

Assessing Young Children in Studies in Early Childhood Education

A very large body of evidence provides support for the effectiveness of early childhood education for children from families with limited economic resources. This evidence has accumulated through many individual evaluations over many years but especially in the last 15 years as more communities and states have implemented and evaluated prekindergarten programs. Early studies, such as the Perry Preschool and the Carolina Abecedarian Projects, found positive short-term findings on assessment constructs but the more notable findings were the long-term effects on significant life outcomes such as high school graduation, earnings, and reduced arrests (Schweinhart et al., 2005). It is doubtful that the program designers ever articulated these long-term outcomes as program goals but the findings nevertheless speak to the power of early childhood programs to change children's trajectories. The results of these and more recent evaluations have been summarized in several thoughtful syntheses, all of which reach the same conclusion: children who attend early childhood programs show better outcomes than those who do not (Melroy, Gardner, & Darling-Hammond, 2019; Phillips, Lipsey, et al., 2017; Yoshikawa, Weiland, & Brooks-Gunn, 2016). The most evidence has been found for literacy and numeracy with somewhat less support for language and social-emotional outcomes. Executive function, which as we noted above is a relatively new outcome area, has only been examined by a few studies to date with some positive findings.

Examining the type of assessments used by the researchers in these studies provides an interesting perspective on how we know what we know about the effectiveness of early education. Table 11.1 presents the assessments from a sample of some studies and evaluations

TABLE 11.1 Measures Used in Selected Studies in Early Childhood Education

Cognitive	*Language*	*Literacy*	*Mathematics*	*Other Academic*	*Social-emotional Behavior*	*Executive Functions*	*Other*
The Perry Preschool Project (Berrueta, Schweinhart, Barnet, Epstein, & Weikart, 1984)							
Stanford Binet Leiter	ITPA, PPVT						
Stanford Binet Leiter WISC	ITPA, PPVT	CAT	CAT		Pupil Behavior Inventory, Ysilanti Rating Scale		
The Carolina Abecedarian Project (Campbell, Helms, Sparling, & Ramey, 1998; Ramey & Campbell, 1991)							
BSID Development, Stanford Binet, WPPSI				PIAT	Infant Behavior Scales (BSID) (adaptive engagement) CBI[T]		
WISC-R		W-J Reading, Written Language, StoCAT	W-J Mathematics CAT	W-J Knowledge	CBI[T] Walker Problem Behavior Identification Checklist CBC[P]		
Head Start Impact Study (Puma, Bell, Cook, & Heid, 2010)							
Leiter	PPVT	Letter Naming Story and Print Concepts Emergent Literacy,[P] PreK CTOPPP Elision, Print Awareness W-J Letter-Word Identification, W-J Spelling, W-J Oral Comprehension W-J Writing Name Task, W-J Word Attack	Counting Bears, W-J Applied Problems, W-J Quantitative Concepts	Color Names W-J Pre-Academic Skills, Academic Skills Report,[T] School Accomplish-ments Report[T]	Child Behavior Problem,[P] Social Competencies Checklist,[P] Social Skills and Positive Approaches to Learning,[P] ASPI[T] Pianta Scale: (Closeness, Conflict, Positive Relationship),[T,P]		McCarthy Scales of Children's Abilities – Draw-a-Design

	PPVT	W-J Letter-Word Identification, W-J Basic Reading Skills, W-J Spelling, W-J Oral Comprehension, W-J Word Attack, Passage W-J Comprehension, W-J Writing Sample; ECLS-K Reading	W-J Applied Problems, W-J Quantitative Concepts, W-J Calculation, Math Reasoning	W-J Academic Skills, W-J Child Self-Assessment, W-J Academic Applications, W-J Academic skills	Pianta Scale: (Closeness, Conflict, Positive Relationship),[T,P] Adapted CPC[P] Strengths and difficulties questionnaire Social Skill and positive approaches to learning, Social competencies, Child Self-Assessment		
Early Head Start Evaluation (Love, Chazan-Cohen, Raikes, & Brooks-Gunn, 2013; Vogel, Xue, Moiduddin, Kisker, & Carlson, 2010)							
BSID	MacArthur CDI,[P] PPVT	W-J Letter-Word ID	W-J Applied Problems		CBCL[P] (aggression) FACES Social Behavior Problems,[P] Play task (child engagement), Leiter-R, BSID (emotion regulation)	Play task (sustained attention) Leiter-R (attention)	FACES Positive Approaches to Learning[P]
WISC IV- Matrix Reasoning	PPVT				CBCL[P] Self description questionnaire (competence, interest in reading, math; competence with peers) Bullying Scale, self report of delinquent behavior		

(Continued)

TABLE 11.1 (Cont.)

Cognitive	*Language*	*Literacy*	*Mathematics*	*Other Academic*	*Social-emotional Behavior*	*Executive Functions*	*Other*
NICHD Study of Early Care and Youth Development (Belsky et al., 2007; NICHD Early Child Care Research Network, 2006; Vandell et al., 2010)							
Bayley Scales of Mental Development, Bracken Concept Scale – school readiness	Preschool Language Scale, Reynell Develop-mental Language Scales, W-J Letter-Word Identification		W-J Applied Problems		Adaptive Social Behavior Inventory,^{P,T} Social Skills Questionnaire,^{P} California Preschool Social Competency Scale,^{T} Child Behavior Checklist,^{P,T} Observation of structured play episodes (peer relations)	W-J Cognitive Memory for Sentences subtest, Continuous Performance Task (sustained attention)	
	W-J Letter Word Identification, W-J Picture Vocabulary, W-J Verbal Analogies, W-J Passage Comprehension		W-J Applied Problems		Social Skills Questionnaire,^{P} Child Behavior Checklist,^{T} Student-Teacher Relationship Scale^{T} (conflict with teacher), Mock Report Cards^{T} (work habits), Teacher Checklist of Peer Relations^{T} (social-emotional functioning); Age 15: Self Interview (risk taking) Weinberger Adjustment Inventory (adapted – impulsivity), Youth Self Report (Externalizing problems)		
National Center for Early Learning and Development (Early et al., 2005)							
	PPVT, OWLS Oral Expression,	W-J Letter-Word Identification, Identifying Letters	W-J Applied Problems, Identifying Numbers		Teacher-Child Rating Scale^{T}		FACES Color Bears
Abbott Preschool Program (Barnett, Jung, Youn, & Frede, 2013; Frede, Jung, Barnet, Lamy, & Figueras, 2007)							
	PPVT	Preschool CTOPPP	W-J Applied Problems				
	Grade 3–5: NJASK Language	Grade 3–5: NJASK Literacy	Grade 3–5: NJASK Math	4th Grade: NJASK Science			

Tennessee Prekindergarten Program (Lipsey, Farran, & Durkin, 2018)							
	W-J Oral Compre-hension, W-J Picture Vocabulary	W-J Letter-Word Identification, W-J Spelling Grades K-3 measures: W-J Passage Comprehension	W-J Applied Problems, W-J Quantitative Concepts Grades K-3 measures: W-J Calculation	ACBR	Cooper-Farran Behavior Rating Scales		
	W-J Oral Comprehension, W-J Picture Vocabulary	W-J Letter-Word Identification, W-J Spelling Grades K-3 measures: W-J Passage Comprehension	W-J Applied Problems, W-J Quantitative Concepts Grades K-3 measures: W-J Calculation				
Tulsa Pre-K Program (Gormley & Gayer, 2005; Gormley, Phillips, & Gayer, 2008)							
ECSI Cognitive	ECSI Language	W-J Spelling, W-J Letter-Word ID,	W-J Applied Problems		ECSI Social-Emotional		ECSI Motor
		W-J Letter-Word Identification, W-J Spelling	W-J Applied Problems				
Boston Public Schools prekindergarten model (Yudron, Weiland, & Sachs, 2016)							
	PPVT	W-J Letter-Word Identification	W-J Applied Problems			FDS, the Pencil Tap, TOQ	

Note: Measures above the dotted line were administered in Prekindergarten or Kindergarten. Measures below the dotted line were administered in Grade 1 or above. Measures were direct assessments unless otherwise noted. ACBR= Academic Classroom and Behavior Record; ASPI= Adjustment Scale for Preschool Intervention; BSID = Bayley Scales of Infant Development; CAT = California Achievement Test; CBI = Classroom Behavior Inventory; CTOPPP = Comprehensive Test of Phonological and Print Processing; ECSI= Early Childhood Skills Inventory; FDS = Forward Digit Span; ITPA = Illinois Test of Psycholinguistic Ability; Leiter = Leiter International Performance Scale; OWLS = Oral PIAT = Peabody Individual Achievement Test; PPVT = Peabody Picture Vocabulary Test; RAPT = Reading Aloud Profile-Together; TEMA-3 = Test of Early Mathematics Ability Third Edition; TOQ = Task Orientation Questionnaire; WISC-R = Wechsler Intelligence Scale for Children – Revised; W-J = Woodcock Johnson Psychoeducational Battery- Third Edition; WPPSI = Wechsler Preschool and Primary Scale of Intelligence. TTeacher or Caregiver completed.

[T]Teacher or Caregiver completed.

[P]Parent completed.

that have looked at the effectiveness question. Some were focused on a particular program (Perry, Head Start, Boston) whereas others looked at programs within a general category (NICHD, NCEDL). Many more studies could have been included in the table but the message about the type and focus of the assessments would likely be similar. Our goal was to place each assessment in the most appropriate construct area given the study information but we readily acknowledge that a more thorough reading of the entire body of work on the study or communication with the associated researchers might have placed some assessments in a different column. The assessments for each study above the dotted line were administered to children in preschool and kindergarten with those below the line given to the children when they were older.

The table reveals both similarities and differences across the studies. Consensus exists at the level with regard to the major outcome areas but there is less commonality at the level of sub-construct and even less at the level of the assessment – although there are some recurring tools such as the PPVT, the McCarthy Scales, and the Woodcock-Johnson. The university research projects such as the Abecedarian and Perry and the federally supported studies and evaluations, i.e., Head Start, NICHD, and NCEDL, tend to have included more assessments than the PreK evaluations, which probably reflects that the former set of studies were better resourced than the latter. As noted above, early childhood studies have moved away from using IQ tests as outcome measures although some studies do describe the PPVT as a measure of general ability. Language, literacy, and math were included as outcome areas in many studies. Approaches to learning and executive function have been included less frequently. Practical and resource constraints dictate limits on the amount of assessment and therefore only a sample of an outcome area's contents are measured. Most studies assess only one or two sub-areas (e.g., letter-word identification, applied problems) and using these to represent the entire outcomes content, which is made up of a far broader set of skills. Evidence for this is provided in the summaries of the findings, which tend to be at the superordinate level (e.g., differences were found in mathematics). The narrowness of the measurement is a serious limitation for understanding impacts and can be especially problematic if a study finds no differences on a sub-area but generalizes that conclusion to the entire outcome area when in fact the study only looked at a very limited sample of content.

Most of the tools are standardized direct assessments with the exception of social-emotional outcomes. For social-emotional constructs, the studies tended to use rating scales completed by a teacher or parent or both. Very few of these studies employed observation-based methods except for the NICHD study, which included observations of play to assess peer relations.

The heavy reliance on direct assessments in this body of research gives rise to an interesting paradox: the studies find effects despite the research pointing to the difficulties of obtaining accurate information on what young children know and can do through direct assessments. Although not included in the table, studies that look at the relationship between a program feature and outcomes also overwhelmingly rely on direct assessments. How can studies find effects with assessments so ill-suited to the nature of the young child? The psychometric data supports the use of direct assessments so is this body of evidence stronger than other evidence that indicates these techniques are a mismatch for the young child?

At the heart of the paradox is the question of the validity of the information produced by the assessments used in these studies. Validity refers to the interpretation of the scores and the use of an assessment for a given purpose (American Educational Research Association et al., 2014). One will often see a particular assessment described as valid but such as description is not appropriate because validity is not a characteristic of an assessment. The same tool could provide valid information for one purpose but not for another. Traditional validity frameworks analyzed the accuracy of an instrument based on its relation to the truest measure of the construct (criterion validity) (American Educational Research Association et al., 2014). This type of validity requires a clear criterion be identified and that data be collected and compared between the instrument being tested and the criterion (National Research Council, 2008). Conceptions of what constitutes validity evidence have evolved over time from single or even multiple types of validity such as criterion, content, and construct validity to a more integrated view of validity. Current thinking conceptualizes validity as an argument that lays out the proposed uses and interpretations of the information produced by the assessment and examines both the reasonableness of the argument and the strength of the evidence that supports it (Kane, 1992; National Research Council, 2008). Validity evidence refers to the evidence that is collected to determine whether an assessment produces information that is suitable for its intended purpose. Reliability evidence, a special type of validity evidence, addresses consistency of the assessment over repeated administrations, assessors, or forms.

The commercially available tools or tools used with the studies in Table 11.1 were not developed for use in research and evaluation. Those who use a tool for a new purpose need to establish the validity of the interpretations of the scores for that purpose (American Educational Research Association et al., 2014). Research in early childhood education does not have a history of constructing validity arguments (Kane, 1992) for the use of particular assessments in studies. Researchers tend to cite the test developers' psychometric information as evidence of validity, but the validity argument needs to reflect the entire set of claims being made for specific interpretation and use for the scores in a given study. The failure to examine validity of the outcomes measures for evaluating programs is not unique to early childhood. In a review of the validity evidence from evaluations of curricular interventions in science and mathematics interventions, Sussman and Wilson (2018) found that most of the validity discussions were inadequate with "fragments of decontextualized evidence" (p. 16), such as reliability coefficients or correlations with other tests, rather than well-developed arguments that used specific evidence to support the interpretations the researchers were putting forward.

A validity claim that early childhood researchers have made implicitly for their study's assessment battery is that the assessments have sufficient sensitivity to detect differences between the program and the comparison groups. The large number of studies that have found positive effects provides support for this claim. The articulation of this claim in the validity argument is important because it also underscores that a failure to find differences could indicate a measurement problem or a true lack of differences. A validity claim that needs to be addressed if studies are to begin to examine questions about differential benefits from early education (see, e.g., Phillips, Johnson et al., 2017) is the assumption that the assessment tools used provide information that is equally accurate for children from different cultural and socioeconomic backgrounds. An example of a validity claim that might be difficult to substantiate in some studies is that assessment content is an accurate reflection of the breadth of the outcome the study seeks to address.

The myriad of issues surrounding direct assessment notwithstanding, the continued pattern of differences favoring the children who received early education suggests that at least some of the tools have been adequate to address questions about effectiveness and thus provide information that has validity for this purpose. The direct assessment paradox is only partly resolved, however, because the current body of work leaves unanswered the question of what could be learned with measurement that was more suited to the nature of the young child and whether current measurement tools are adequate to address second-generation questions about what works for whom.

Perplexing Issues

The nature of the assessments used to measure outcomes may be a contributing factor to two issues that have perplexed researchers in early childhood evaluations. One of these is the phenomenon of "fadeout" that has been observed in several studies of the long-term early childhood programs. As the program children and the comparison children move through elementary school, the early skills advantage for the program group disappears. Both groups of children continue to gain skills but their achievement levels converge (Yoshikawa et al., 2016). This "catch-up" or "fadeout" phenomenon has been extensively discussed with various explanations for its occurrence (Ansari & Pianta, 2018; Bailey et al., 2017; Brooks-Gunn, 2011; Duncan & Magnuson, 2013; Haskins, 2016; Jenkins et al., 2018). Some programs like the Perry Preschool Project do not see a skills advantage in school but find differences in later life outcomes (Schweinhart et al., 2005). The Head Start Impact Study found scores converging in elementary school but did not follow children into adulthood to see if differences reappear (Puma et al., 2012). Other studies using different methodologies, however, have found long-term effects for Head Start (Carneiro & Ginja, 2014; Deming, 2009; Garces, Thomas, & Currie, 2002). The evidence on the long-term impact of PreK programs on academic achievement in school is mixed across studies with some studies finding differences that continue through elementary school and others showing convergence (Melroy et al., 2019; Phillips, Lipsey et al., 2017).

Among the multiple factors that could be contributing to these conflicting findings are the outcomes that were measured and the tools used to measure them. In the case of the Perry Preschool Project, it is highly likely that the program and control children differed in critical ways during their school years given the differences in life outcomes that were found later. If the wrong outcomes were assessed in elementary school, however, these differences would have gone undetected. The differences could have been in non-cognitive skills such as personality traits, goals, character, motivation, and preferences for success in life (Kautz, Heckman, Diris, ter Weel, & Borghans, 2015). Some studies measure constrained skills, which could explain why the groups' skills converge. Also, some studies rely on standardized achievement tests as the measure of academic achievement in elementary school. Different achievement tests are not comparable to one another in what they measure. Equally important, standardized achievement tests typically measure a fairly narrow set of skills and not more complex skills such as higher order thinking skills (Conley & Darling-Hammond, 2013; National Research Council, 2001; Sussman & Wilson, 2018). Gains for the early education group would appear to not be maintained if preK programs are impacting later cognitive or academic skills but not the skills measured by standardized achievement tests, i.e., correct construct but not the correct sub-construct.

Another perplexing issue comes from studies of child care. Researchers have been puzzled by the recurring failure to find a stronger relationship between program quality and child outcomes. Early research was concerned with the possible negative effects of child care (Belsky & Steinberg, 1978) but later research has focused on the relationship between quality (a program feature in Figure 11.1) and outcomes. A large body of research on the impact of environment on development as well as knowledge of what constituted high-quality care would suggest that child outcomes would improve as the quality of the program increased. The general finding is that program quality measures show a modest relationship with child outcomes measures (Burchinal, 2018; Burchinal, Kainz, & Cai, 2011; Zaslow et al., 2011). The direction is as expected but the relationship contains some noise; it is not as clear nor as strong as would be expected. Understanding the relationship between program quality and child outcomes has important policy implications because a number of states are using child outcomes assessments as part of validating their Quality Rating and Improvement Systems (QRIS) (Hestenes et al., 2015; Sabol & Pianta, 2015; Soliday Hong, Howes, Marcella, Zucker, & Huang, 2015). Several hypotheses have been advanced for the moderate relationship including that program quality is not well specified, not well measured, or both; that the relationship is tied to a threshold of quality; that the research does not consider the extent of exposure to differing levels of quality; and that the conceptualization or measurement of the child outcomes is inadequate (Brunsek et al., 2017; Burchinal, 2018; Hatfield, Burchinal, Pianta, & Sideris, 2016; Zaslow et al., 2011). Although far more of the discussion has focused on the measurement of program quality, issues with the child outcomes measures may warrant more consideration. One of the implied validity claims is that the assessments have sufficient sensitivity to address this research question. Given the variety of outcome measures used and the multiple potential threats to validity when assessing young children, especially young from diverse backgrounds, this level of accuracy might be beyond the capacity of the assessments that have been used in these studies.

Conclusion: A Need for Better Assessments

The assessments used in research on early childhood education have both evolved and stayed the same over the last 70 years. The constructs used in studying the effects of early education have moved away from IQ, which dominated measurement in the 1960s. Many studies now embrace the elusive but nevertheless widely accepted construct of school readiness as the intended outcome of early childhood education. School readiness is typically conceptualized as consisting of multiple construct areas including language, early literacy, mathematics, social-emotional skills, and, more recently, processes supported by neuroscience research such as executive function. Although the constructs have changed over time, how we assess outcomes has changed very little. The majority of the assessments used in studies of early childhood programs involve a stranger removing the child from everyday routines and asking the child to perform decontextualized tasks on command. Researchers and professional associations alike have criticized this type of assessment, noting that it is inconsistent with the nature of the young child, especially those from non-majority cultures, and unlikely to produce good information. Bronfenbrenner's characterization of developmental psychology in 1977 applies equally well to assessment in early childhood education studies today: it is the science of the strange behavior of children in strange situations with strange adults for the briefest possible periods of time (Bronfenbrenner, 1977).

Despite a heavy reliance on assessment methods ill-suited to young children, several decades of research have produced a substantial body of knowledge demonstrating the benefits of early education. Early education programs have been shown to improve outcomes across a variety of program types including state prekindergarten, district programs, Head Start, and child care (Melroy et al., 2019; Phillips, Lipsey et al., 2017). The consistency of the findings across so many studies despite possible validity issues with the assessment results for some children can be interpreted as strong support for the power of early education. Some of these studies are structurally similar to an hourglass in that the program goals are broad (e.g., prepare children to succeed in school), the assessments are narrow (e.g., a tiny sample of behavior in a few outcome areas), and study conclusions are as broad as the program goals (e.g., improved school readiness). Hopefully, as the field moves forward to answer more sophisticated and nuanced questions about specific features of early education and sub-groups for children, studies will reflect better alignment between program goals, constructs, assessments, and conclusions.

The call for improved assessment techniques for use with young children has been repeated many times over the years (National Research Council, 2000, 2008; Shephard, 1994). In 1995, the technical work group to the National Goals Panel recommended abandoning testing strategies that rely on single administrations, are uni-dimensional, and fail to capture cultural variability. They called for data collection strategies that include input from teachers, families, and children themselves and for methods that extract from information from common or everyday experiences (Kagan et al., 1995). In 2000, the Committee on Integrating the Science of Early Childhood Development concluded: "Thus, for more than three decades, researchers and service providers have struggled with both the identification of significant child outcomes and their valid and reliable measurement" (Shonkoff & Phillips, 2000, p. 347). Two decades later, the struggle continues. The Committee went on to call for improved tools to measure self-regulation, emotional development, the capacity to make friends and engage with others, language use (as contrasted with vocabulary), and executive functions. Progress has been made in some of these areas, most notably self-regulation and executive function, but other areas such as emotional development continue to be a challenge. Hirsh-Pasek et al. (2005) advocated for an assessment approach that examines skills as they are used in every situation and that recognizes the interactions among outcome areas. They argued that integrative assessment techniques that capture the nature of children's learning and evaluate how competencies in different developmental domains interact are more consistent with the nature of the young child and better able to address validity concerns around culture and context. Much more research on early childhood assessment is needed, especially around issues of measuring behavior in context and appropriately for all children including children from non-mainstream cultures, dual language learners, and children with disabilities. Innovative use of technology might hold promise as a mechanism for capturing children's everyday behaviors through videotaping and electronic toys or tablets that can record how children naturally interact with them. New technologies also could support efficient transcriptions and video coding, which would allow these techniques to be feasibly used on a large scale.

In addition to continuing to explore what should be measured and how to measure it, more research is needed on the meaning of assessment scores for young children. A question that merits more discussion in the science of early childhood education is "how does one interpret the performance or behavior of a young child at one point in time?" Research has established that development is not linear; that the traditional domains are not distinct; that there is a great deal of variation in the age at which children acquire different skills; that what the child does not know today, he or she can (and should) learn tomorrow; that correlation is not causation, and so even if a study has shown an association between an early assessment score and a later score that should not mean that children are glued to a particular trajectory throughout their school years. Many years ago, one of us conducted a study that compared children who had attended Head Start in a school district with those who were wait listed. The study followed the children through high school graduation and found that the Head Start children outperformed the wait-listed children on several outcomes measures – but also that the overall performance of the Head Start children was not very good at each of the follow up points (Hebbeler, 1985). The general reaction to these findings within the district was that Head Start had failed. My response to that was that the school system had these children for 13 additional years after Head Start. The correlations between their Grade 3 and Grade 11 test scores were distressingly high. Multiple data points along the way showed that these children needed help but nothing happened to change their trajectories. In this particular context, an assessment at age 8 became a measure of stable trait rather than a malleable indicator. Similarly, studies of the long-term effects of early education bank on the perverse hope that no future teacher of the comparison children will ever be able to help them achieve what they missed by not attending preschool.

Even for an important outcome area assessed in a meaningful context with culturally sensitive techniques, the question of how to interpret a score that is a snapshot at a single point in time remains and the issue does not change when hundreds of scores are aggregated. That score represents a moment in time for a young child who is developing and learning rapidly and will most certainly perform differently a few months later; or, if assessed on a bad day, would perform differently tomorrow. When assessment scores for young children who perform poorly do not predict their future performance, is that a measurement concern or a cause for celebration? Research must continue using assessments as they are the tools with which we document effectiveness and learn what constitutes good early education for young children. Science would be well served if more studies acknowledge the limitations of our current assessment methods, if new assessment methods are developed that are more suited to young children, and if interpretations more adequately reflect the complexity of development and learning both during and after early childhood.

References

Ackerman, D.J., & Friedman-Kraus, A.H. (2017). *Preschoolers' executive function: Importance, contributors, research needs and assessment options*. Princeton, NJ: Wiley.

American Educational Research Association, American Psychological Association, & National Council on Measurement in Education. (2014). *Standards for educational and psychological testing*. Washington, DC: American Educational Research Association.

Ansari, A., & Pianta, R.C. (2018). The role of elementary school quality in the persistence of preschool effects. *Children and Youth Services Review*, 86, 120–127. doi:10.1016/j.childyouth.2018.01.025

Atkins-Burnett, S. (2007). *Measuring children's progress from preschool through third grade*. Washington, DC: Mathematica Policy Institute.

Bagnato, S.J. (2007). *Authentic assessment for early childhood intervention: Best practices*. New York, NY: Guilford Press. Retrieved from http://www.loc.gov/catdir/enhancements/fy0708/2007006524-t.html

Bailey, D., Duncan, G.J., Odgers, C.L., & Yu, W. (2017). Persistence and fadeout in the impacts of child and adolescent interventions. *Journal of Research on Educational Effectiveness*, 10(1), 7–39. doi:10.1080/19345747.2016.1232459

Barnett, W.S., Jung, K., Youn, M., & Frede, E.C. (2013). *Abbott preschool program longitudinal effects study: Fifth grade follow-up*. New Brunswick, NJ: National Institute for Early Education Research.

Belsky, J., & Steinberg, L.D. (1978). The effects of day care: A critical review. *Child Development*, 49(4), 929–949. doi:10.2307/1128732

Belsky, J., Vandell, D.L., Burchinal, M., Clarke-Stewart, K.A., McCartney, K., Owen, M.T., & The NICHD Early Child Care Research Network. (2007). Are there long-term effects of early child care? *Child Development*, 78(2), 681–701. doi:10.1111/j.1467-8624.2007.01021.x

Berrueta, J.R., Schweinhart, L.J., Barnet, W.S., Epstein, A., & Weikart, D.P. (1984). *Changed lives: The effects of the Perry Preschool Program on youths through age 19*. Ysiplanti, MI: High/Scope Educational Research Foundation.

Bierman, K.L., Nix, R.L., Greenberg, M.T., Blair, C., & Domitrovich, C.E. (2008). Executive functions and school readiness intervention: Impact, moderation, and mediation in the Head Start REDI program. *Development and Psychopathology*, 20(3), 821–843. doi:10.1017/S0954579408000394

Bishop-Josef, S.J., & Zigler, E. (2011). The cognitive/academic emphasis versus the whole child approach: The 50-year debate. In E. Zigler, W.S. Gilliam, & W.S. Barnett (Eds.), *The Pre-K debates: Current controversies and issues* (pp. 83–88). Baltimore, MD: Brooks.

Blair, C., Knipe, H., Cummings, E., Baker, D.P., Gamson, D., Eslinger, P., & Thorne, S.L. (2007). A developmental neuroscience approach to the study of school readiness. In R.C. Pianta, M.J. Cox, &

K. Snow (Eds.), *School readiness and the transition to kindergarten in the age of accountability* (pp. 149–174). Baltimore, MD: Brooks.

Bodrova, E., & Leong, D.J. (2018). Common assessment terms and how to use them: A glossary for early educators. In H. Bohart & R. Procopio (Eds.), *Spotlight on young children: Observation and assessment*. Washington, DC: National Association for the Education of Young Children.

Bronfenbrenner, U. (1977). Toward an experimental ecology of human development. *American Psychologist* (July), 513–531.

Brooks-Gunn, J. (2011). Early childhood education: The likelihood of sustained effects. In E. Zigler, W.S. Gilliam, & W.S. Barnet (Eds.), *The PreK debates: Current issues and controversies* (pp. 200–217). Baltimore, MD: Brooks.

Brunsek, A., Perlman, M., Falenchuk, O., McMullen, E., Fletcher, B., & Shah, P.S. (2017). The relationship between the Early Childhood Environment Rating Scale and its revised form and child outcomes: A systematic review and meta-analysis. *PLOS ONE*, 12(6), e0178512. doi:10.1371/journal.pone.0178512

Burchinal, M. (2018). Measuring early care and education quality. *Child Development Perspectives*, 12(1), 3–9. doi:10.1111/cdep.12260

Burchinal, M., Kainz, K., & Cai, Y. (2011). How well do our measures of quality predict child outcomes? In M. Zaslow, I. Martinez-Beck, K. Tout, & T. Halle (Eds.), *Quality measurement in early childhood settings* (pp. 11–31). Baltimore, MD: Paul H. Brookes Publishing.

Cabell, S.Q., Justice, L.M., Zucker, T.A., & Kilday, C.R. (2009). Validity of teacher report for assessing the emergent literacy skills of at-risk preschoolers. *Language, Speech, and Hearing Services in Schools*, 40 (2), 161–173.

Campbell, F.A., Helms, R., Sparling, J.J., & Ramey, C.T. (1998). Early childhood programs and success in school. In W.S. Barnett & S.S. Boocock (Eds.), *Early care and education for children in poverty: Promises, programs, and long-term results* (pp. 145–166). Albany, NY: State University of New York Press.

Campbell, S.B., Denham, S.A., Howarth, G.Z., Jones, S.M., Whittaker, J.V., Williford, A.P., … Darling-Churchill, K. (2016). Commentary on the review of measures of early childhood social and emotional development: Conceptualization, critique, and recommendations. *Journal of Applied Developmental Psychology*, 45, 19–41. doi:10.1016/j.appdev.2016.01.008

Carneiro, P., & Ginja, R. (2014). Long-term impacts of compensatory preschool on health and behavior: Evidence from Head Start. *American Economic Journal: Economic Policy*, 6(4), 135–173. doi:10.1257/pol.6.4.135

Conley, D.T., & Darling-Hammond, L. (2013). *Creating systems of assessment for deeper learning*. Stanford, CA: Stanford Center for Opportunity Policy in Education.

Cross, C.T., Woods, T.A., & Schweingruber, H. (Eds.). (2009). *Mathematics learning in early childhood: Paths toward excellence and equity*. Washington, DC: National Academies Press.

Darling-Churchill, K.E., & Lippman, L. (2016). Early childhood social and emotional development: Advancing the field of measurement. *Journal of Applied Developmental Psychology*, 45, 1–7. doi:10.1016/j.appdev.2016.02.002

Darling, L., & Grace, C. (2015). Using observational assessment in conducting research with young children in classroom settings. In O. Saracho (Ed.), *Handbook of Research Methods in Early Childhood Education: Review of research methodologies* (vol. 2). Charlotte, NC: Information Age Publishing.

Deming, D. (2009). Early childhood intervention and life-cycle skill development: Evidence from Head Start. *American Economic Journal: Applied Economics*, 1(3), 111–134. doi:10.1257/app.1.3.111

Diamond, A. (2013). Executive functions. *Annual Review of Psychology*, 64, 135–168. doi:10.1146/annurev-psych-113011-143750

Division for Early Childhood of the Council for Exceptional Children. (2007). Position statement – Promoting positive outcomes for children with disabilities: Recommendations for curriculum, assessment, and program evaluation. Retrieved from: https://www.decdocs.org/position-statement-promoting-positi

Downer, J.T., Booren, L.M., Lima, O.K., Luckner, A.E., & Pianta, R.C. (2010). The Individualized Classroom Assessment Scoring System (inCLASS): Preliminary reliability and validity of a system for observing preschoolers' competence in classroom interactions. *Early Childhood Research Quarterly*, 25 (1), 1–16. doi:10.1016/j.ecresq.2009. 08. 004

Duncan, G., & Magnuson, K. (2013). Investing in preschool programs. *Journal of Economic Perspectives*, 27 (2), 109–132.

Early, D., Barbarin, O., Bryant, D., Burchinal, M., Chang, F., Clifford, R., ... Barnett, W.S. (2005). *Pre-kindergarten in eleven states: NCEDL's Multi-state study of pre-kindergarten & study of State-Wide Early Education Programs (SWEEP): Preliminary descriptive report* (NCEDL Working Paper). Chapel Hill, NC: NCEDL.

Epstein, A., Schweinhart, L.J., & DeBruin-Parecki, A. (2004). Assessing children's development: strategies that complement testing. In D. Koralek (Ed.), *Spotlight on young children and assessment* (pp. 45–52). Washington, DC: National Association for the Education of Young Children.

Frede, E.C., Jung, K., Barnet, W.S., Lamy, C.E., & Figueras, A. (2007). *The Abbott Preschool Program Longitudinal Effects Study (APPLES)*. New Brunswick, NJ: National Institute for Early Education Research. Retrieved from https://www.researchgate.net/profile/William_Barnett6/publication/237135385_THE_ABBOTT_PRESCHOOL_PROGRAM_LONGITUDINAL_EFFECTS_STUDY_APPLES/links/0f31752df25721883f000000.pdf

Garces, E., Thomas, D., & Currie, J. (2002). Longer-term effects of Head Start. *American Economic Review*, 92(4), 999–1012. doi:10.1257/00028280260344560

Gormley, W.T., Jr., & Gayer, T. (2005). Promoting school readiness in Oklahoma: An evaluation of Tulsa's pre-k program. *The Journal of Human Resources*, 40(3), 533–558.

Gormley, W.T., Jr., Phillips, D., & Gayer, T. (2008). The early years: Preschool programs can boost school readiness. *Science*, 320(5884), 1723–1724.

Graue, M.E. (1992). Social interpretations of readiness for kindergarten. *Early Childhood Research Quarterly*, 7, 225–243.

Halle, T.G., & Darling-Churchill, K.E. (2016). Review of measures of social and emotional development. *Journal of Applied Developmental Psychology*, 45, 8–18. doi:10.1016/j.appdev.2016.02.003

Haskins, R. (2016). American policy on early childhood education and development: Many programs, great hopes, modest impacts. *Behavioral Science & Policy*, 2(1), 1–8.

Hatfield, B.E., Burchinal, M.R., Pianta, R.C., & Sideris, J. (2016). Thresholds in the association between quality of teacher–child interactions and preschool children's school readiness skills. *Early Childhood Research Quarterly*, 36, 561–571. doi:10.1016/j.ecresq.2015.09.005

Hebbeler, K. (1985). An old and new question on the effects of early education for children from low-income families. *Education Evaluation and Policy Analysis*, 7(3), 204–216.

Hernandez, M.W., Demir-Lira, O.E., Graf, E., Rohn, M., & Smith, K.E. (2016). *Effective early childhood and pre-kindergarten programs, interventions & best practices: A review of literature in the domains of oral language & literacy, math and social-emotional development*. Chicago, IL: University of Chicago.

Hestenes, L.L., Kintner-Duffy, V., Wang, Y.C., La Paro, K., Mims, S.U., Crosby, D., ... Cassidy, D.J. (2015). Comparisons among quality measures in child care settings: Understanding the use of multiple measures in North Carolina's QRIS and their links to social-emotional development in preschool children. *Early Childhood Research Quarterly*, 30, 199–214. doi:10.1016/j.ecresq.2014.06.003

Hirsh-Pasek, K., Kochanoff, A., Newcombe, N.S., & de Villiers, J. (2005). Using scientific knowledge to inform preschool assessment: Making the case for "empirical validity". Vol. XIX. SRCD Social Policy Report: SRCD.

Jenkins, J.M., Watts, T.W., Magnuson, K., Gershoff, E.T., Clements, D.H., Sarama, J., & Duncan, G.J. (2018). Do high-quality kindergarten and first-grade classrooms mitigate preschool fadeout? *Journal of Research on Educational Effectiveness*, 11(3), 339–374. doi:10.1080/19345747.2018.1441347

Jones, S.M., Zaslow, M., Darling-Churchill, K.E., & Halle, T.G. (2016). Assessing early childhood social and emotional development: Key conceptual and measurement issues. *Journal of Applied Developmental Psychology*, 45, 42–48. doi:10.1016/j.appdev.2016.02.008

Kagan, S.L., & Kauerz, K. (2007). Reaching for the whole: Integration and alignment in early education policy. In R.C. Pianta, M.J. Cox, & K.L. Snow (Eds.), *School readiness and the transition to kindergarten in the era of accountability* (pp. 11–30). Baltimore, MD: Paul H. Brookes Publishing.

Kagan, S.L., Moore, E., & Bredekamp, S. (Eds.). (1995). *Reconsidering children's early development and learning: Toward common views and vocabulary*. Washington, DC: Government Printing Office: Report of the National Education Goals Panel, Goal 1 Technical Planning Group.

Kagan, S.L., Rosenkotter, S., & Cohen, N. (Eds.). (1997). *Considering child-based results for young children: Definitions, desirability, feasibility, and next steps*. New Haven, CT: Yale Bush Center on Child Development and Social Policy.
Kane, M.T. (1992). An argument-based approach to validity. *Psychological Bulletin*, 112, 527–535.
Kautz, T., Heckman, J.J., Diris, R., ter Weel, B., & Borghans, L. (2015). *Fostering and measuring skills: Improving cognitive and non-cognitive skills to promote lifetime success*. Cambridge, MA: National Bureau of Economic Research. Retrieved from http://www.nber.org/papers/w20749
Konold, T.R., & Pianta, R.C. (2007). The influence of informants on ratings of children's behavioral functioning: A latent variable approach. *Journal of Psychoeducational Assessment*, 25(3), 222–236. doi:10.1177/0734282906297784
La Paro, K.M., & Pianta, R.C. (2000). Predicting children's competence in the early school years: A Meta-analytic review. *Review of Educational Research*, 70(4), 443–484.
Lambert, R.G., Kim, D.H., & Burts, D.C. (2014). Using teacher ratings to track the growth and development of young children using the Teaching Strategies GOLD® assessment system. *Journal of Psychoeducational Assessment*, 32(1), 27–39.
Lazar, I., Darlington, R., Murray, H., Royce, J., Snipper, A., & Ramey, C.T. (1982). Lasting effects of early education: A report from the consortium for longitudinal studies. *Monographs of the Society for Research in Child Development*, 47(2/3), 1–151. doi:10.2307/1165938
Lipsey, M.W., Farran, D.C., & Durkin, K. (2018). Effects of the Tennessee Prekindergarten Program on children's achievement and behavior through third grade. *Early Childhood Research Quarterly*. doi:10.1016/j.ecresq.2018.03.005
Love, J.M. (2001, December). Instrumentation for state readiness assessment: Issues in measuring children's early development and learning. Paper presented at the Assessing the State of State Assessments Symposium, SERVE, Atlanta, GA.
Love, J.M., Chazan-Cohen, R., Raikes, H., & Brooks-Gunn, J. (2013). What makes a difference: Early Head Start evaluation findings in a developmental context. *Monographs of the Society for Research in Child Development*, 78(1), vii–viii, 1–173.
Marcella, J., & Howes, C. (2015). Using naturalistic observation to study children's engagement within early childhood classrooms: Review of research methodologies. In O. Saracho (Ed.), *Handbook of research methods in early childhood education* (vol. II, pp. 25–53). Charlotte, NC: Information Age Publishing.
McLean, M., Wolery, M., & Bailey, D.B., Jr. (2004). *Assessing infants and preschoolers with special needs* (3rd ed.). Upper Saddle River, NJ: Pearson.
Meisels, S. (2007). Accountability in early childhood: No easy answers. In R.C. Pianta, M.J. Cox, & K. Snow (Eds.), *Schools readiness and the transition to kindergarten in the era of accountability* (pp. 31–47). Baltimore, MD: Brookes.
Meisels, S.J. (1996). *New visions for the developmental assessment of infants and young children*. Washington, DC: Zero to Three.
Meisels, S.J., & Atkins-Burnett, S. (2000). The elements of early childhood assessment. In J. P. Shonkoff & S.J. Meisels (Eds.), *Handbook of early childhood intervention* (2nd ed., pp. 231–257). New York, NY: Cambridge University Press.
Meisels, S.J., Bickel, D.D., Nicholson, J., Xue, Y., & Atkins-Burnett, S. (2001). Trusting teachers' judgments: A validity study of a curriculum-embedded performance assessment in kindergarten to grade 3. *American Educational Research Journal*, 38(1), 73–95.
Melroy, B., Gardner, M.K., & Darling-Hammond, L. (2019). *Untangling the evidence on preschool effectiveness: Insights for policymakers*. Palo Alto, CA: Learning Policy Institute. Retrieved from https://learningpolicyinstitute.org/product/untangling-evidence-preschool-effectiveness-report
Morrison, F., & Grammer, J.K. (2016). Executive function in preschool-age children: Integrating measurement, neurodevelopment, and translational research. In J.A. Griffin, P. McCardle, & L.S. Freund (Eds.). Washington, DC: American Psychological Association.
Nagle, R.J. (2007). Issues in preschool assessment. In B. Bracken & R.J. Nagle (Eds.), *Psychoeducational assessment of preschool children* (pp. 29–48). Mahwah, NJ: Lawrence Erlbaum.
NAEYC & NAECS/SDE. (2003). Early childhood curriculum, assessment, and program evaluation. Building an effective, accountable system in programs for children birth through 8. Retrieved from:

https://www.naeyc.org/sites/default/files/globally-shared/downloads/PDFs/resources/position-statements/pscape.pdf

NASP. (2015). Early childhood services: Promoting positive outcomes for young children (Position statement). Bethesda, MD: Author.

National Early Childhood Accountability Task Force. (2007). *Taking stock: Assessing and improving early childhood learning and program quality*. Pew Charitable Trusts.

National Education Goals Panel. (1991a). *Goal 1 Technical Planning Subgroup report on school readiness*. Washington, DC: Author.

National Education Goals Panel. (1991b). *The national education goals report*. Washington, DC: Author.

National Institute of Child Health and Human Development. (2002). *Early Childhood education and school readiness: Conceptual models, constructs and measures*. Washington, DC: Author.

National Research Council. (2000). *Eager to learn: Educating our preschoolers*. Washington, DC: The National Academies Press. Retrieved from https://www.nap.edu/catalog/9745/eager-to-learn-educating-our-preschoolers

National Research Council. (2001). *Knowing what students know: The science and design of educational assessment*. Washington, DC: The National Academies Press.

National Research Council. (2008). *Early childhood assessment: Why, what, and how?*Washington, DC: The National Academies Press.

Neisworth, J.T., & Bagnato, S.J. (2004). The mismeasure of young children. *Infants and Young Children*, 17(3), 198–212.

NICHD Early Child Care Research Network. (2006). Child-care effect sizes for the NICHD Study of early child care and youth development. *American Psychologist*, 61(2), 99–116.

Paris, S.G. (2005). Reinterpreting the development of reading skills. *Reading Research Quarterly*, 40(2), 184–202.

Perry, N.E., & Meisels, S. (1996). *How accurate are teacher judgments of students' academic performance?* Washington, DC: National Center for Educational Statistics.

Phillips, D.A., Johnson, A.D., Weiland, C., & Hutchison, J.E. (2017). *Public preschool in a more diverse America: Implications for next-generation evaluation research*. Ann Arbor, MI: Poverty Solutions at the University of Michigan.

Phillips, D.A., Lipsey, M.W., Dodge, K.A., Haskins, R., Bassok, D., Burchinal, M., ... Weiland, C. (2017). *Puzzling it out: The current state of scientific knowledge pre-kindergarten effects: A consensus statement*. Washington, DC: Brookings Institution.

Puma, M., Bell, S., Cook, R., & Heid, C. (2010). *Head Start Impact Study Final Report*. Washington, DC: Administration for Children and Families, U.S. Department of Health and Human Services. Retrieved from https://www.acf.hhs.gov/opre/resource/head-start-impact-study-final-report-executive-summary

Puma, M., Bell, S., Cook, R., Heid, C., Broene, P., Jenkins, F., ... Downer, J. (2012). *Third grade follow-up to the Head Start Impact Study Final Report* (OPRE Report 2012–2045). Washington, DC: U.S. Department of Health and Human Services.

Ramey, C.T., & Campbell, F.A. (1991). Poverty, early childhood education, and academic competence: The Abecedarian experiment. In A.C. Huston (Ed.), *Children in poverty* (pp. 190–221). New York, NY: Cambridge University Press.

Ramey, C.T., & Ramey, S.L. (2004). Early educational interventions and intelligence: Implications for Head Start. In E. Zigler & S. Styfco (Eds.), *The Head Start debates* (pp. 3–17). Baltimore, MD: Brooks.

Raver, C. (2002). Emotions matter: Making the case for the role of young children's emotional development for early school readiness. *Social Policy Report*, 16(3), 1–20.

Raver, C., & Blair, C. (2016). Neuroscientific insights: Attention, working memory, and inhibitory control. *The Future of Children*, 26(2), 95–118.

Reynolds, A.J., Richardson, B.A., Hayakawa, M., Lease, E.M., Warner-Richter, M., Englund, M. M., ... Sullivan, M. (2014). Association of a full-day vs part-day preschool intervention with school readiness, attendance, and parent involvement. *Journal of the American Medical Association*, 312(20), 2126–2134. doi:10.1001/jama.2014.15376

Russo, J.M., Williford, A.P., Markowitz, A.J., Vitiello, V.E., & Bassok, D. (2019). Examining the validity of a widely-used school readiness assessment: Implications for teachers and early childhood programs. *Early Childhood Research Quarterly*, 28(48), 14–25.

Sabol, T.J., & Pianta, R.C. (2015). Validating Virginia's quality rating and improvement system among state-funded pre-kindergarten programs. *Early Childhood Research Quarterly*, 30, 183–198. doi:10.1016/j.ecresq.2014.03.004

Schorr, L.B. (1997). Judging interventions by their results. In L.S. Kagan, S.E. Rosenkoetter, & N. Cohen (Eds.), *Considering child-based results for young children: Definitions, desirability, feasibility, and next steps* (pp. 36–47). New Haven, CT: Yale Bush Center in Child Development and Social Policy.

Schrag, R.D.A., Styfco, S., & Zigler, E. (2004). Familiar concepts, new name: Social competence/school readiness. In E. Zigler & S. Styfco (Eds.), *The Head Start debates* (pp. 19–25). Baltimore, MD: Brooks.

Schweinhart, L.J., Montie, J., Xiang, Z., Barnett, W.S., Belfield, C.R., & Nores, M. (2005). *Lifetime effects: The High/Scope Perry Preschool Study through age 40.* Ypsilanti, MI: High/Scope Press.

Schweinhart, L.J., & Weikart, D.P. (1981). Effects of the Perry Preschool Program on youths through age 15. *Journal of the Division for Early Childhood*, 4(1), 29–39.

Scott-Little, C., Kagan, S.L., & Frelow, V.S. (2006). Conceptualization of readiness and the content of early learning standards: The intersection of policy and research? *Early Childhood Research Quarterly*, 21(2), 153–173. doi:10.1016/j.ecresq.2006.04.003

Scott-Little, C., Kagan, S.L., Frelow, V.S., & Reid, J. (2009). Infant-toddler early learning guidelines: The content that states have addressed and implications for programs serving children with disabilities. *Infants & Young Children*, 22(2), 87–99 doi:10.1097/IYC.1090b1013e3181a1002f1094b

Sekino, Y., & Fantuzzo, J. (2005). Validity of the Child Observation Record: An investigation of the relationship between COR dimensions and social-emotional and cognitive outcomes for Head Start children. *Journal of Psychoeducational Assessment*, 23(3), 242–261.

Shephard, L. (1994). The challenges of assessing young children appropriately. *Phi Delta Kappan*, 76, 206–212.

Shephard, L., Kagan, S., & Wurtz, E. (1998). *Principles and recommendations for early childhood assessments.* Washington, DC: National Education Goals Panel.

Shonkoff, J., & Phillips, D. (Eds.). (2000). *From neurons to neighborhoods: The science of early childhood development.* Washington, DC: National Academy Press.

Shores, E.F., & Grace, C. (1998). *The portfolio book: A step-by-step guide for teachers.* Lewisville, NC: Gryphon House.

Snow, C., & Oh, S.S. (2011). Assessment in early literacy research. In S.B. Neuman & D.K. Dickinson (Eds.), *Handbook of early literacy research, Volume 3* (pp. 375–395). New York, NY: Guilford Press.

Snow, C.E., & Matthews, T.J. (2016). Reading and language in the early grades. *Future of Children*, 26 (2), 57–74.

Snow, K. (2007). Integrative views of the domains of child function: Unifying school readiness. In R. Pianta, M.J. Cox, & K. Snow (Eds.), *School readiness and the transition to kindergarten in the era of accountability* (pp. 197–216). Baltimore, MD: Brookes.

Soderberg, J., Stull, S., Cummings, K., Nolan, E., & McCutchen, D.G. (2013). *Inter-rater reliability and concurrent validity study of the Washington Kindergarten Inventory of Developing Skills (WaKIDS).* Seattle, WA: University of Washington.

Soliday Hong, S.L., Howes, C., Marcella, J., Zucker, E., & Huang, Y. (2015). Quality Rating and Improvement Systems: Validation of a local implementation in LA County and children's school-readiness. *Early Childhood Research Quarterly*, 30, 227–240. doi:10.1016/j.ecresq.2014.05.001

Suen, H.K., Logan, C.R., Neisworth, J.T., & Bagnato, S.J. (1995). Parent-professional congruence: Is it necessary? *Journal of Early Intervention*, 19(3), 243–252.

Sussman, J., & Wilson, M.R. (2018). The use and validity of standardized achievement tests for evaluating new curricular interventions in mathematics and science. *American Journal of Evaluation*, 40(2), 190–213. doi:10.1177/1098214018767313

Urban, M., & Swadener, B.B. (2016). Democratic accountability and contextualized systemic evaluation: A comment on the OECD initiative to launch an International Early Learning Study (IELS). *International Critical Childhood Policy Studies Journal*, 5(1), 6–18.

Vandell, D.L., Belsky, J., Burchinal, M., Steinberg, L., Vandergrift, N., & NICHD Early Child Care Research Network. (2010). Do effects of early child care extend to age 15 years? Results from the NICHD study of early child care and youth development. *Child Development*, 81(3), 737–756. doi:10.1111/j.1467-8624.2010.01431.x

Vogel, C.A., Xue, Y., Moiduddin, E.M., Kisker, E.E., & Carlson, B.L. (2010). *Early Head Start children in grade 5: Long-term follow-up of the Early Head Start research and evaluation study sample*Washington, DC: Office of Planning, Research, and Evaluation, Administration for Children and Families, U.S. Department of Health and Human Services.

Waterman, C., McDermott, P.A., Fantuzzo, J., & Gadsden, V. (2012). The matter of assessor variance in early childhood education. Or whose score is it anyway? *Early Childhood Research Quarterly*, 27, 46–54.

Whitehurst, G.J., & Lonigan, C.J. (1998). Child development and emergent literacy. *Child Development*, 69(3), 848–872. doi:10.1111/j.1467-8624.1998.tb06247.x

Yoshikawa, H., Weiland, C., & Brooks-Gunn, J. (2016). When does preschool matter? *The Future of Children*, 26(2), 21–35.

Yudron, M., Weiland, C., & Sachs, J. (2016). *BPS K1DS: Piloting the Boston Public Schools' pre-kindergarten model in community-based organizations*. Boston, MA: Boston Public Schools.

Zaslow, M., Anderson, R., Redd, Z., Wessel, J., Turillo, L., & Burchinal, M. (2011). *Quality, dosage, thresholds, and features in early childhood settings: A review of the literature*. Washington, DC: OPRE. Retrieved from https://eric.ed.gov/?id=ED579878

Zigler, E., Styfco, S., & Gilman, E. (1993). The national Head Start Program for disadvantaged preschoolers. In E. Zigler & S. Styfco (Eds.), *Head Start and beyond: A national plan for extended early childhood intervention* (pp. 1–41). New Haven, CT: Yale University Press.

Zigler, E., & Styfco, S.J. (2000). Pioneering steps (and fumbles) in developing a federal preschool intervention. *Topics in Early Childhood Special Education*, 20(2), 67–70.

Zigler, E.F., & Trickett, P.K. (1978). IQ, social competence, and the evaluation of early childhood intervention programs. *American Psychologist*, 33, 789–798.

12

MEDICAL SCIENCES INFLUENCES ON EARLY CHILDHOOD EDUCATION

Magdalena Janus, Ayesha Siddiqua and Salmi Noor

Early childhood is a very special period in a child's life. For both the child and family, it can be magical, full of new discoveries and progress every day, with developmental advances happening at a much faster pace than at any other period. Children acquire the developmental building blocks that allow them to further hone their cognitive and social and emotional abilities past this stage. There is a common understanding now, supported by much evidence, that many of the ailments experienced by people in later childhood and in adulthood have their roots in early development, through a process that has been called a "biological embedding" (Hertzman & Boyce, 2010). This term has been coined to illustrate how disadvantage, experienced in early life, can influence a child's biology and physical health, and then, in addition to social determinants, result in heightened susceptibility to disease through adulthood. For example, small changes in the nutritional status in fetal perinatal stages, especially if combined with disadvantageous environments, can disrupt metabolic mechanisms enough to increase risk for a variety of diseases, long after the conditions have improved (Hanson & Gluckman, 2008; Hertzman & Boyce, 2010).

Medical sciences, defined broadly to include clinical as well as contributing basic science, have had a profound impact on the understanding of early childhood development and its life-long consequences. This knowledge has been translated into early interventions with children themselves, their families, and to some extent early childhood education through a number of both direct and indirect pathways, some of which are the subjects of other chapters. In this chapter, we will outline these pathways and describe to a fuller extent the ones that are not considered elsewhere.

It may not come as a surprise that the impact of medical sciences is mostly visible in the early education and care of children with special medical needs. While we have come a long way since the times when children with physical and developmental health problems were institutionalized without access to early – or any – education, the challenges that medical conditions bring to typical early educational contexts continue. In some respects, the challenges reflect the need to balance the inclusive education laws with the specialized attention that many children with extensive developmental delays require (Wolery, 1999). In others, it is the attitude that may be a barrier – when children's disability is seen as preventing them from being capable of otherwise healthy development (Rosenbaum, 2007). They may also reflect a lack of logistic and system capacity that prevents children from accessing early education and accommodation resources (Fowler, Schwartz, & Atwater, 1991). More and more of what used to be perceived as "difficult" behavior is recognized as a diagnosable condition, and support services are linked to a diagnosis. However, health systems need to be capable of bringing about the functional needs assessment early enough to facilitate the earliest possible development of individually adjusted education practices, and that is often not the case (Wolery, 1999; Janus, Lefort, Cameron, & Kopechanski, 2007). Older, as well as more recent, research (and practice) indicates that waiting lists for early assessments are growing. Only a fraction of children whose difficulties are observed in early settings receive appropriate accommodation and individualized education at school entry (Janus, Kopechanski, Cameron, & Hughes, 2008).

Perhaps the strongest and most comprehensive impact of medical science on the knowledge of child development, and the resulting applied field of early education, happened through neuroscience and a constantly growing refinement of the understanding of brain development. The intersection of neuroscience and early education is the subject of Chapter 13.

Another pathway through which medicine influenced early education was providing the foundations for the ideas and practice focused on remedying the learning needs of children with non-typical development – that is, special education. While now we usually think of educational settings providing special education, it certainly originated from medicine and medical institutions (Odom et al., 2005). Finally, expansion of the field of public health into realms of early development has brought advancements in health, safety, and hygiene for all children.

In this chapter, we will focus largely on, first, the impact of medicine on the field of special early education and the growing use of the International Classification of Functioning that illustrates merging of intervention and rehabilitation processes into early education. Second, we will briefly address the impact of public health on early education through its emphasis on prevention.

Special Education

Medicine's direct impact on early education in general is minimal in comparison with all the other disciplines reviewed in this book. However, it certainly provided support to the origins of special education – including that of special early education: physicians were the professionals who provided recognition of the needs of young children with disabilities, as well as designed services for their education (Odom et al., 2005). Such education followed the tradition of medical care as it originally occurred in residential facilities for children with physical and mental disabilities. Experimental approaches to preschool education of children with developmental delays, referred to at that time as mental retardation, have provided original insights (Kirk, 1958). With the growth in contribution from other sciences – psychology, sociology, anthropology, and ethnography (Odom et al., 2005; Underwood, Valeo, & Wood, 2012), the medical model became less relevant to early childhood education, and more specifically aligned with particular intervention and preventive strategies characteristic of specific disabilities. Since medical advancements have been making it increasingly possible for many children to survive and overcome (at least to some degree) perinatal health risks, the discipline is now also contributing to the growth of a population that has more complex educational needs than others.

The medical model of education is based on the assumption that children's non-typical learning needs are a direct result of a health disorder (Massoumeh & Leila, 2012). It has had a significant influence on defining the landscape for special early education. It has been used both for the diagnosis and educational treatment of children with a variety of disabilities. Under the medical model, disability is understood as an impairment and inclusion is perceived as an assimilation process, where educators are expected to "fit" the child into the early childhood education settings (Mackenzie, Cologon, & Fenech, 2016). The emphasis of this model is to change the child rather than the environment or the pedagogy. This model proposes that a comprehensive diagnosis of physical, neurological, or behavioral disorders should precede the interventions provided in educational settings (Massoumeh & Leila, 2012). Most early intervention programs used in the context of early childhood education are designed based on the medical model of disability, where pathological characteristics of children are determined, followed by the implementation of interventions that seek to change the characteristics of the children to improve their performance (Underwood et al., 2012). As many less severe disorders are difficult to diagnose in very early childhood, this model has been criticized for imposing constraints on how well young children with disabilities can be supported both at home and in early education settings.

With shifts in attitudes and practices regarding the education of children with disabilities, there has been a move towards advocating for inclusive learning settings (Shonkoff & Meisels, 2000). The medical model has significant influence over training methods for teachers, which are based on a biological-psychological perspective and highlight the importance of teaching procedures that promote appropriate sensory stimulation and perceptual organization (Massoumeh & Leila, 2012). Consistent with the medical model, a defining feature of inclusive early childhood education programs is the presence of various therapists who provide services that are integrated in the child's activities within the classroom routine (Case-Smith & Holland, 2009). In order to facilitate frequent practice and reinforcement of a new skill, occupational therapists, physical therapists, and speech and language therapists work with children with disabilities in their natural learning settings. Case-Smith and Holland (2009) argue that the therapist's role is to support teachers and their learning needs in relation to each particular child, as much as it is to develop interventions for children. As the educators' lack of comfort with providing special education has also been noted, presumably to ultimately ensure improvements in teacher education, this argument indirectly suggests the growing need for improved preparation of early educators in their training.

Teachers are generally not taught special education unless they participate in additional training after receiving their teaching degree. Those teachers who do take special education courses are more willing to accept children with special needs into their classrooms and are also more confident in their ability to teach special education (Stephens & Braun, 1980). Teachers who have experienced more disability-related training perceive fewer needs and barriers to inclusion of atypical children (Mulvihill, Shearer, & Van Horn, 2002). Similarly, previous experience with atypically developing children may also help to alleviate teacher concerns or perceived barriers. Mahoney and Wheeden (1999) showed that teaching style has an impact on the learning of children with special needs. Teaching styles which are best described as directive create a higher likelihood that atypically developing children will pay attention to the teacher, rather than engage in play. On the other hand, when teaching style is highly responsive, and non-directive, children are more likely to initiate activities on their own, but do not usually include the teacher in these interactions. Unfortunately, we have found no evidence of whether such findings are incorporated into the training of teachers.

Although the medical model has played a pivotal role in shaping special early education, there have been several critiques of this approach. There is a lack of consensus regarding whether most learning disorders are caused by biological factors only, as there is evidence indicating that interactions between the environment and the biological systems also account for disabling conditions (Massoumeh & Leila, 2012). Furthermore, there is a general lack of evidence supporting the notion that methods based on the medical model are effective in achieving educational goals. Given these limitations, a new version of the medical model has been proposed. This new model postulates that most effective interventions should involve a combination of medical and behavioral approaches to support optimal classroom learning for children with disabilities that takes into account the environment in which children are situated (Forness & Kavale, 2001a).

Despite these criticisms, special early education programs based on the medical model continue to prevail. An example of this approach can be observed for children with autism spectrum disorders (ASD), where the medical model is used to individualize instructions and interventions for these children in learning settings. This individualized approach to education begins with a variety of diagnostic or norm-referenced assessments that provide information about the strengths and weaknesses of children with ASD in comparison with typically developing children (Barton, Lawrence, & Deurloo, 2012). Results of these assessments play an integral role in developing an individualized education program for children with ASD, which may involve adapting the classroom environment to suit the children's needs (Barton et al., 2012). It is widely recognized that no single program fits all, and diverse interventions have been utilized to support early education for children with ASD. Some interventions have been designed to address broad issues such as communication training, structured teaching, and behavioral programming, whereas other interventions are targeted to more specific issues such as social skill development, academic instruction, as well as development of life skills (Bryson, Rogers, & Fombonne, 2003).

The medical model has been particularly influential in informing special early education programs for children with attention deficit hyperactivity disorder (ADHD). Traditionally, a behavioral model has been emphasized to support learning of children with ADHD. This involves the common implementation of psycho-social, school-based interventions, which focus on children's behavioral manifestations without taking into account any underlying issues such as genetic factors, neurobiologic functioning, or specific psychiatric diagnosis (Forness & Kavale, 2001a). However, research emerging in recent years has highlighted the need to shift from psychodynamic approaches and begin to consider medical interventions, such as psychopharmacologic treatments, given their potential to support the academic and social functioning of children with ADHD. There is evidence showing that, among children with ADHD, adding comprehensive behavioral intervention to medical treatment has not significantly improved gains made from using medical treatment alone, suggesting the critical role the medical model can play in informing special education decisions (Forness & Kavale, 2001b). A high rate of stimulant medication use among children with ADHD has been observed in special education programs (Schnoes, Reid, Wagner, & Marder, 2006). This example illustrates the limitations in applying any one specific model to early education without considering the specifics of a health disorder.

Finding a space for special early education and supported learning in a regular preschool classroom has been assisted by medical professionals through the rising importance of developmental and behavioral pediatrics, a branch of pediatrics that emerged in the mid-twentieth century and settled as a subspecialty some 20–30 years later, which is meant to focus on psychological, social, and learning problems of children and adolescents (Haggerty & Friedman, 2003). The developmental pediatricians themselves are still shy of delineating the precise characteristics of their field – but they are often on the front lines in helping children with developmental learning difficulties (Garner et al., 2012). While now, in essence, developmental and behavioral pediatrics merges developmental psychology and psychiatry with medicine, it is really seen as the medical specialty that needs to address the common – yet severe enough to seek intervention – problems of children's behavior and learning. In Canada, more so than in the United States, it is also the complementary specialization, child psychiatry, in whose purview lie more severe mental health impairments, that helps set the child and family on a path towards resources that would guarantee support in classrooms. In many communities, developmental pediatricians and child psychiatrists are powerful advocates for the inclusion of children with disabilities in early education settings and for integrated provision of educational supports (e.g., Affiliated Services for Children and Youth in Hamilton, Canada, https://ascy.ca/about-ascy/). These efforts are certainly changing both the physical space and the practice of early education.

International Classification of Function of Children and Youth

In 2007, the WHO International Classification of Function (ICF) was extended, to include specific ways in which children function, into the ICF-CY for Children and Youth (Simeonsson et al., 2003; World Health Organization, 2007). This comprehensive framework for classifying the functioning of children with disabilities has been used successfully, though perhaps not as widely as it should be, to guide the education of children with functional disabilities.

The innovation of the ICF-CY, which should make it particularly attractive to early education settings, is that it combines the medical model of disability, which focuses on bodily impairments, and the social model, which proposes that participation restrictions occur due to society's construction of difficulties (Maxwell, Alves, & Granlund, 2012). Although the theoretical model presented by the ICF-CY has the potential to inform education systems, it remains infrequently used, likely because of the systemic silos in which care for young children with disability typically occur, such as health, education, and social services (Janus et al., 2007). The ICF-CY framework's future potential also lies in the opportunity to provide a common language to connect such different systems, especially in early childhood, but also over a child's growth from early to later education (Kraus de Camargo, 2011).

The ICF-CY's application in the intersection of developmental pediatrics and education would facilitate identifying children on the basis of their functional profiles rather than diagnostic labels. This would ensure that intervention planning does address functional needs, which is often not the case when only broad diagnoses are used through the predominant categorical identification approach (Simeonsson et al., 2003). Moreover, the participation component of the ICF-CY is particularly relevant for early education, as it focuses on the availability and access to everyday activities, which has important implications for participating in an educational setting (Maxwell et al., 2012). The environment is a contextual component of the ICF-CY, which can be used to examine the barriers and facilitators for participation (Rosenbaum, 2007). Several studies in educational research have examined participation in relation to environmental dimensions emphasized in the ICF-CY, as well as used this framework to evaluate the congruency between the assessment-intervention processes for children with special needs (Maxwell et al., 2012; Castro, Pinto, & Simeonsson, 2012). Thus far, there are few examples of special education programs designed based on direct consultation of this framework. A notable one, however, is the development of ICF-based procedures for educational practices in Switzerland (Hollenweger, 2011).

Public Health

Contributions of public health, epidemiology, and developmental pediatrics have also been reflected in the structural aspects of early childhood education settings. From updated health and safety regulations, to procedures minimizing infection and increasing health standards, the advances in medicine have improved the quality of the places where the youngest children receive their early education.

The past few decades have seen a rise in the use of child-care services, largely due to a changing workplace landscape, and an increase in mobility (Holmes, Morrow, & Pickering, 1996; Willer, 1991). Working mothers found it increasingly difficult to find family or relatives living in close proximity to provide child care, and therefore formal child-care settings grew in popularity. Currently, the majority of children under 5 attend some form of early child care (Alkon, To, Mackie, Wolff, & Bernzweig, 2010). This has impacted the epidemiology of communicable and infectious diseases as children in child care are more likely to be exposed to them than are their peers who are cared for at home (Goodman, Osterholm, Granoff, & Pickering, 1984; Lu et al., 2004). Characteristics such as age, sex, immunologic status, and length of time since enrolment in the early childhood education setting are all associated with risks of illness, where respiratory tract infections and bacterial diseases were found to be most common (Holmes et al., 1996). Increase in these illnesses among children in early education settings has resulted in overuse, and sometimes misuse, of antibiotics and other medications (Reves & Jones, 1990). Due to the higher frequency of antibiotic prescriptions to treat infectious diseases in children attending center-based early education, the emergence of antibiotic-resistant organisms has been a growing public health concern. Various public health and regulatory bodies collaborated to establish methods to reduce antimicrobial resistance and continued safety in child-care settings, with proper hand-washing standards, disinfection guidelines, and effective child immunization practices making up some of the top recommendations (*Caring for our children: National health and safety performance standards: Guidelines for out-of-home child care*, American Academy of Pediatrics, American Public Health Association, & National Resource Center for Health and Safety in Child Care, 2002). Despite establishment of these guidelines, early childhood educators and directors express the inadequacy of the existing health and safety protocols, and feel that proper training by certified healthcare professionals is also needed to complement the more technical aspects of the safety standards of early child-care settings (Alkon et al., 2010).

Early childhood education providers also express that there is a lack of proper procedures and standards surrounding injury prevention, specifically for outdoor play (Story, Kaphingst, & French, 2006). This is concerning as safe and nurturing play environments have been shown to be crucial for a child's long-term developmental and physical health. In 2017, The Consumer Product Safety Commission reported more than 72,000 injuries to children aged 0–4 years from child day care equipment alone (*Data Highlights*, National Electronic Injury Surveillance System, 2017). The most common form of injury in early childhood education settings as a result of this equipment was falls (Kotch, Chalmers, Langley, & Marshall, 1993), and when the causes for falls were further assessed, results pointed to the poor maintenance and the height of playground equipment as well as the improper surfaces below it (Briss, Sacks, Addiss, Kresnow, & O'Neil, 1995; Kotch et al., 1993; Waibel & Misra, 2003). Over time, improvements in policy on safer playground materials have contributed to extensive literature aimed towards fostering both creative and protective play environments for children in school and early child-care settings (Howard et al., 2005; Mott et al., 1997).

In the context of increased demand for early child care and education, as well as recognition of the importance of high-quality early learning for children's future health and development, the process of addressing health and safety in early child-care settings generated attention that is being translated into improvements. For example, in the United States this amplified attention contributed to the development of safety checklists that incorporate national health and safety standards in early childhood education settings to provide stimulating and positive learning environments for all children (Alkon, To, Wolff, Mackie, & Bernzweig, 2008). Health practitioners propose training nurses as child-care consultants who visit early education centers at regular intervals to help directors identify gaps in health and safety protocols, and develop plans to address these gaps through shared decision making (Gaines, Wold, Spencer, & Leary, 2005; Lucarelli, 2002). There are also suggestions that employing child-care consultants with expert training in nursing or medicine can help achieve safe and healthy environments for children in their early years (Cianciolo, Trueblood-Noll, & Allingham, 2004). Despite considerable progress in improving the safety of center-based early education, it still remains to be seen which health professionals are best suited to act as child-care consultants in early education settings, and how efficient the health and safety checklists are in comparison with other valid safety measures (Alkon et al., 2008; Gaines et al., 2005).

Conclusion

In conclusion, medicine by its very nature is aimed at solving problems – those associated with health. It may even be fair to say that, historically, early education has not been within the purview of medical sciences, except where it concerned special educational needs of children with medical conditions. For much of our history, children whose needs diverged from typically developing children were rarely welcome in mainstream classrooms. When early education, such as preschools, became more widespread, the issue of supporting learning of children with medically compromising conditions in early education settings needed to be addressed.

More recently, the medical profession, with pediatricians in developed countries especially, has taken the social determinants of health approach to child development (Commission on Social Determinants of Health, 2008). This is leading to promotion of early education as a factor demonstrated to have a protective effect for children's future well-being, and thus leading to healthier life trajectories by simply ensuring that there are enough opportunities for early education for all children (Burger, 2010; Moore, McDonald, Carlon, & O'Rourke, 2015; Allensworth, 2011). Some initiatives (such as Early Child Development Taskforce of the Global Partnership for Children with Disabilities, 2014) are specifically focused on young children with disabilities and their families.

Medicine, through its many branches, such as pediatrics, epidemiology, and public health, has had a historical impact on the practice of early education over the last century. While this impact is changing, it remains ever present. The recent efforts led by many medical professionals to view young children with disabilities as children first, and to ensure that they are fully present in all early education initiatives globally, might yet prove to be the most lasting and powerful influence medicine has had on the field and science of early childhood education and society as a whole.

References

Alkon, A., To, K., Mackie, J.F., Wolff, M., & Bernzweig, J. (2010). Health and safety needs in early care and education programs: What do directors, child health records, and national standards tell us? *Public Health Nurse*, 27(1), 3–16. doi:10.1111/j.1525-1446.2009.00821.x

Alkon, A., To, K., Wolff, M., Mackie, J.F., & Bernzweig, J. (2008). Assessing health and safety in early care and education programs: Development of the CCHP health and safety checklist. *Journal of Pediatric Health Care*, 22(6), 368–377. doi:10.1016/j.pedhc.2007. 11. 002

Allensworth, D.D. (2011). Addressing the social determinants of health of children and youth: a role for SOPHE members. *Health Education & Behaviour*, 38(4), 331–338. doi:10.1177/1090198111417709

American Academy of Pediatrics, American Public Health Association, & National Resource Center for Health and Safety in Child Care. (2002). *Caring for our children: National health and safety performance standards: Guidelines for out-of-home child care*. American Academy of Pediatrics. Retrieved from https://www.researchconnections.org/childcare/resources/1859

Barton, E.E., Lawrence, K., & Deurloo, F. (2012). Individualizing interventions for young children with autism in preschool. *Journal of Autism and Developmental Disorders*, 42(6), 1205–1217. doi:10.1007/s10803-011-1195-z

Briss, P.A., Sacks, J.J., Addiss, D.G., Kresnow, M.J., & O'Neil, J. (1995). Injuries from falls on playgrounds. Effects of day care center regulation and enforcement. *Archives of Pediatrics and Adolescent Medicine*, 149(8), 906–911.

Bryson, S.E., Rogers, S.J., & Fombonne, E. (2003). Autism spectrum disorders: early detection, intervention, education, and psychopharmacological management. *Canadian Journal of Psychiatry*, 48(8), 506–516. doi:10.1177/070674370304800802

Burger, K. (2010). How does early childhood care and education affect cognitive development? An international review of the effects of early interventions for children from different social backgrounds. *Early Childhood Research Quarterly*, 25, 140–165. doi:10.1016/j.ecresq.2009. 11. 001

Case-Smith, J., & Holland, T. (2009). Making decisions about service delivery in early childhood programs. *Language, Speech, and Hearing Services in Schools*, 40, 416–423. doi:0161-1461/09/4004doi:0416

Castro, S., Pinto, A., & Simeonsson, R.J. (2012). Content analysis of Portuguese individualized education programmes for young children with autism using the ICF-CY framework. *European Early Childhood Education Research Journal*, 22(1), 91–104. doi:10.1080/1350293X.2012.704303

Cianciolo, S., Trueblood-Noll, R., & Allingham, P. (2004). Health consultation in early childhood settings. *YC Young Children*, 59(2), 56–61.

Commission on Social Determinants of Health. (2008). Closing the gap in a generation: Health equity through action on the social determinants of health. Final Report of the Commission on Social Determinants of Health. Retrieved from http://apps.who.int/iris/bitstream/handle/10665/43943/9789241563703_eng.pdf?sequence=1

Early Child Development Taskforce of the Global Partnership for Children with Disabilities. (2014). Concept note & work plan. Retrieved from http://www.gpcwd.org/uploads/2/6/0/9/26097656/ecdtf_concept_note__work_plan_cnwp_1.14.pdf

Forness, S.R., & Kavale, K.A. (2001a). ADHD and a return to the medical model of special education. *Education and Treatment of Children*, 24(3), 224–247.

Forness, S.R., & Kavale, K.A. (2001b). Ignoring the odds: Hazards of not adding the new medical model to special education decisions. *Behavioural Disorders*, 26(4), 269–281.

Fowler, S.A., Schwartz, I., & Atwater, J. (1991). Perspectives on the transition from preschool to kindergarten for children with disabilities and their families. *Exceptional Children*, 58(2), 136–145.

Gaines, S.K., Wold, J.L., Spencer, L., & Leary, J.M. (2005). Assessing the need for child-care health consultants. *Public Health Nurse*, 22(1), 8–16. doi:10.1111/j.0737-1209.2005.22103.x

Garner, A.S., Shonkoff, J.P., Siegel, B.S., Dobbins, M.I., Earls, M.F., Garner, A., …Wood, D.L. (2012). Early childhood adversity, toxic stress, and the role of the pediatrician: Translating developmental science into lifelong health. *American Academy of Pediatrics*, 129(1), e224–e231. doi:10.1542/peds.2011-2662

Goodman, R.A., Osterholm, M.T., Granoff, D.M., & Pickering, L.K. (1984). Infectious diseases and child day care. *Pediatrics*, 74(1), 134–139.

Haggerty, R.J., & Friedman, S.B. (2003). History of developmental-behavioral pediatrics. *Journal of Developmental and Behavioural Pediatrics*, 24(1 Suppl), S1–18.

Hanson, M.A., & Gluckman, P.D. (2008). Developmental origins of health and disease: New insights. *Basic and Clinical Pharmacology and Toxicology*, 102(2), 90–93. doi:10.1111/j.1742-7843.2007.00186.x

Hertzman, C., & Boyce, T. (2010). How experience gets under the skin to create gradients in developmental health. *Annual Review of Public Health*, 31, 329–347. doi:10.1146/annurev.publhealth.012809.103538

Hollenweger, J. (2011). Development of an ICF-based eligibility procedure for education in Switzerland. *BMC Public Health*, 11 (Suppl 4), S7. doi:10.1186/1471–2458–11-s4-s7

Holmes, S.J., Morrow, A.L., & Pickering, L.K. (1996). Child-care practices: Effects of social change on the epidemiology of infectious diseases and antibiotic resistance. *Epidemiological Review*, 18(1), 10–28.

Howard, J.S., Sparkman, C.R., Cohen, H.G., Green, G., & Stanislaw, H. (2005). A comparison of intensive behavior analytic and eclectic treatments for young children with autism. *Research in Developmental Disabilities*, 26(4), 359–383.

Janus, M., Kopechanski, L., Cameron, R., & Hughes, D. (2008). In transition: Experiences of parents of children with special needs at school entry. *Early Childhood Education Journal*, 35(5), 479–485. doi:10.1007/s10643-10007-0217-0

Janus, M., Lefort, J., Cameron, R., & Kopechanski, L. (2007). Starting kindergarten: Transition issues for children with special needs. *Canadian Journal of Education*, 30(3), 628–648.

Kirk, S.A. (1958). *Early education of the mentally retarded: An experimental study*. Illinois: University of Illinois Press.

Kraus de Camargo, O. (2011). Systems of care: Transition from the bio-psycho-social perspective of the International Classification of Functioning, Disability and Health. *Child: Care, Health, and Development*, 37(6), 792–799. doi:10.1111/j.1365–2214.2011.01323.x

Kotch, J.B., Chalmers, D., Langley, J.D., & Marshall, S.W. (1993). Child day care and home injuries involving playground equipment. *Journal of Paediatrics and Child Health*, 29(3), 222–227.

Lu, N., Samuels, M., Shi, L., Baker, S., Glover, S., & Sanders, J. (2004). Child day care risks of common infectious diseases revisited. *Child: Care, Health and Development*, 30(4), 361–368.

Lucarelli, P. (2002). Raising the bar for health and safety in child care. *Pediatric Nursing*, 28(3), 239–241.

Mackenzie, M., Cologon, K., & Fenech, M. (2016). "Embracing everybody": Approaching the inclusive early childhood education of a child labelled with autism from a social relational understanding of disability. *Australasian Journal of Early Childhood*, 41(2), 4–12.

Mahoney, G., & Wheeden, C.A. (1999). The effect of teacher style on interactive engagement of preschool-aged children with special learning needs. *Early Childhood Research Quarterly*, 14(1), 51–68.

Massoumeh, Z., & Leila, J. (2012). An investigation of medical model and special education methods. *Social and Behavioural Sciences*, 46, 5802–5804.

Maxwell, G., Alves, I., & Granlund, M. (2012). Participation and environmental aspects in education and the ICF and the ICF-CY: Findings from a systematic literature review. *Developmental Neurorehabilitation*, 15(1), 63–78. doi:10.3109/17518423.2011.633108

Moore, T.G., McDonald, M., Carlon, L., & O'Rourke, K. (2015). Early childhood development and the social determinants of health inequities. *Health Promotion International*, 30 (Suppl 2), ii102–115. doi:10.1093/heapro/dav031

Mott, A., Rolfe, K., James, R., Evans, R., Kemp, A., Dunstan, F., … Sibert, J. (1997). Safety of surfaces and equipment for children in playgrounds. *The Lancet*, 349(9069), 1874–1876.

Mulvihill, B.A., Shearer, D., & Van Horn, M.L. (2002). Training, experience, and child care providers' perceptions of inclusion. *Early Childhood Research Quarterly*, 17(2), 197–215. doi:10.1016/S0885-2006(02)00145-X

National Electronic Injury Surveillance System. (2017). NEISS Data Highlights – 2017. Retrieved from https://www.cpsc.gov/s3fs-public/2017-Neiss-data-highlights.pdf?3i3POG9cN.rIyu2ggrsUkD1XU_zoiFRP

Odom, S.L., Brantlinger, E., Gersten, R., Horner, R.H., Thomson, B., & Harris, K.R. (2005). Research in special education: Scientific methods and evidence-based practices. *Exceptional Children*, 71(2), 137–148.

Reves, R., & Jones, J. (1990). Antibiotic use and resistance patterns in day care centers. *Seminars in Pediatric Infectious Diseases*, 1, 212–221.

Rosenbaum, P. (2007). The environment and childhood disability: opportunities to expand our horizons. *Developmental Medicine and Child Neurology*, 49(9), 643. doi:10.1111/j.1469-8749.2007.00643.x

Schnoes, C., Reid, R., Wagner, M., & Marder, C. (2006). ADHD among students receiving special education services: A national survey. *Exceptional Children*, 72(4), 483–496.

Shonkoff, J.P., & Meisels, S.J. (2000). *Handbook of Early Childhood Intervention*. Cambridge: Cambridge University Press.

Simeonsson, R.J., Leonardi, M., Lollar, D., Bjorck-Akesson, E., Hollenweger, J., & Martinuzzi, A. (2003). Applying the International Classification of Functioning, Disability and Health (ICF) to measure childhood disability. *Disability and Rehabilitation*, 25(11–12), 602–610. doi:10.1080/0963828031000137117

Stephens, T.M., & Braun, B.L. (1980). Measures of regular classroom teachers' attitudes toward handicapped children. *Exceptional Children*, 46(4), 292–294.

Story, M., Kaphingst, K.M., & French, S. (2006). The role of child care settings in obesity prevention. *The Future of Children*, 16(1), 143–168.

Underwood, K., Valeo, A., & Wood, R. (2012). Understanding inclusive early childhood education: A capability approach. *Contemporary Issues in Early Childhood*, 13(4), 290–299.

Waibel, R., & Misra, R. (2003). Injuries to preschool children and infection control practices in childcare programs. *Journal of School Health*, 73(5), 167–172.

Willer, B. (1991). The demand and supply of child care in 1990: Joint findings from the National Child Care Survey 1990 and a profile of child care settings. Retrieved from https://eric.ed.gov/?id=ED339522

Wolery, M. (1999). Children with disabilities in early elementary school. In R.C. Pianta & M.J. Cox (Eds.), *The transition to kindergarten* (pp. 253–280). Baltimore, MD: Paul H. Brookes.

World Health Organization. (2007). International Classification of Functioning, Disability and Health: Children & Youth version. Retrieved from http://apps.who.int/iris/bitstream/handle/10665/43737/9789241547321_eng.pdf?sequence=1

13

THE INFLUENCE OF NEUROSCIENCE ON EARLY CHILDHOOD EDUCATION

Rebecca Distefano, Ellen Galinsky and Philip David Zelazo

Introduction

It is only within very recent history – the past 25 to 30 years – that neuroscience has become a force in child development and educational research, as the tools to study the brain in action have improved and become more readily available. A notable moment in this history was the publication of *Starting Points: Meeting the Needs of Our Youngest Children* in 1994 by the Carnegie Corporation of New York (Carnegie Task Force, 1994). In that report, the authors wrote:

> With the help of powerful new research tools, including sophisticated brain scans, scientists have studied the developing brain in greater detail than ever before.

Research on brain development has progressed considerably since 1994, but even at that time, the following conclusions were drawn:

- First, the brain development that takes place during the prenatal period and in the first year of life is more rapid and extensive than we previously realized.
- Second, brain development is much more vulnerable to environmental influence than we ever suspected.
- Third, the influence of early environment on brain development is long lasting.
- Fourth, the environment affects not only the number of brain cells and number of connections among them, but also the way these connections are "wired."
- And fifth, we have new scientific evidence for the negative impact of early stress on brain function.

Although there was only one other time that the word "brain" appeared in the entire 41-page *Starting Points* report, brain research dominated the extensive media coverage of the report following its release, prompting the Carnegie Corporation of New York and the Families and Work Institute to convene a conference at the University of Chicago in 1996 called "Brain Development in Young Children: New Frontiers for Research, Policy and Practice." That conference brought together leading researchers in neuroscience and in child development because, for the most part, they didn't know each other's research. Its purpose was to determine where the findings of neuroscience and child development research converged and where they differed and to set a research and policy agenda for the future.

The findings of the Chicago Conference were written up in a report called *Rethinking the Brain: New Insights into Early Development* (Shore, 1997), published by the Families and Work Institute and released by President and Mrs. Clinton at a conference in April 1997 entitled, "The White House Conference on Early Childhood Development and Learning." The wide dissemination of *Rethinking the Brain* as well as the White House conference (which was broadcast to nearly 100 locations), and a 1997 special edition of Newsweek Magazine on "Your Child – From Birth to Three" launched the subject of the brain development of young children into educational circles as well as to the general public. Three years later, the 2000 publication of *Neurons to Neighborhoods* by the National Research Council and the subsequent important work of the Center on the Developing Child at Harvard University made knowledge of children's brain development a fundamental and complementary part of understanding the development and learning of young children.

A Historical Perspective: The Work of Montessori, Vygotsky, and Luria

Before contemporary brain scanning tools – like functional magnetic resonance imaging (fMRI) or magnetoencephalography (MEG) – there were a number of leading educators and psychologists whose work foreshadowed the findings of contemporary developmental neuroscience and their implications for education and growth. Key among them were Maria Montessori, Lev Vygotsky, and Alexander Luria.

In her seminal book, *The Montessori Method*, Montessori (1912) noted that children have an inherent motivation to explore their environments and learn from them, and she recognized in this motivation children's inherent drive to develop and grow. In contrast to more didactic pedagogical approaches, Montessori took a more autonomy-supportive approach, and designed an environment filled with opportunities that supported and stimulated children's perceptual, motor, cognitive, social, and emotional development, and encouraged children's autonomy and sense of agency by allowing them to select their own learning challenges based on their interests. Contemporary research on the impact of Montessori education has revealed the extent to which it promotes the healthy development not only of autonomy and agency but of the self-regulatory, executive function skills that underlie them (e.g., Lillard & Else-Quest, 2006).

Vygotsky (1929) also recognized the key role that the environment, especially the cultural context, plays in the development of children's self-regulation. Together with Luria, and building on Pavlov's notion of a second signal system that allowed signs (e.g., words) to signal (or control) a primary signal system, Vygotsky outlined a socio-cultural model of development in which children become more self-regulated based on interpersonal interaction and the internalization of cultural practices, such as speech. According to Vygotsky (1934/1962), the internalization of cultural tools, particularly speech, provided the impetus for qualitative changes in the structure of thought. Whereas speech initially serves a communicative purpose, it subsequently acquires a directive or self-regulatory function that allows children to organize and plan their behavior, essentially rendering them capable of intentional, goal-directed thought and action. A key finding from their research is that, with age, children are able to use increasingly complex verbal plans to guide their behavior (Luria, 1961). The work of Vygotsky and Luria provided an important foundation for contemporary research on the development of executive function skills and their dependence on brain networks involving prefrontal cortex.

Early Intervention Studies: The Perry Preschool Project and Abecedarian

Two comprehensive early childhood interventions – the Perry Preschool Project, begun in the 1960s, and the Abecedarian Project, begun in the 1970s – demonstrated the impact of high-quality early childhood programs on children deemed at risk of subsequent school failure, and in both cases there is evidence that the beneficial effects were brought about in large part because these interventions supported the development of executive function skills. These programs have had a lasting impact because:

- They had sound research designs of randomly assigning children either to an experimental group that participated in the high-quality early childhood intervention or to a control group.
- They were longitudinal, following the children from when they entered the program up through adulthood.
- They used numerous indicators of life success, including school achievement, placement in special education, grades or grade retention, high school completion rates, rates of adult crime and delinquency, employment and earnings as adults, and use of social service programs.

By every indicator, the children in these intervention programs have fared better than the control groups, yielding conservative benefit/cost ratios for every dollar invested of $7.12 for The Perry Preschool Project (Heckman, Moon, Pinto, Savelyev, & Yavitz, 2010) and $7.31 for Abecedarian (Ramey, 2018). That is, over the course of a lifetime, each dollar invested in these programs returns at least 7 dollars back to society. Although these interventions were not informed by neuroscience as it exists today, they were based on many intervention principles and best practices that have emerged in the literature, including the following:

- They began early and were comprehensive – at 6 weeks through kindergarten entry for Abecedarian and 3- to 4-years-old for the Perry Preschool Project. As Jack Shonkoff of the Center on the Developing Child at Harvard University said in a report conducted on the impact of these interventions:

  > These interventions provided positive learning experiences and supportive, growth-promoting environments at a time when the children's brain circuits were being built. Thus, they promoted the development of sturdy brain architecture that provided a stronger foundation for later achievement … It's better to get it right the first time than to try to fix it later.
  >
  > *(Galinsky, 2006, p. 7)*

- Their curricular approaches were based on promoting engaged learning. As Larry Schweinhart of HighScope said, the Perry Curriculum was "not to directly instruct the children, but to support their own choices and to extend those choices" (Galinsky, 2006, p. 8). Their maxim was to have the children plan, do, and review, and for teachers to scaffold children's learning and promote refection on what was learned.

- They took a whole child approach, engaging the child socially, emotionally, physically, and cognitively, and they promoted the learning of academic content as well as skill learning in playful ways (Ramey, 2018). Craig and Sharon Ramey of the Abecedarian Project outline the following program features they consider essential for the success of the Abecedarian Project (Ramey & Ramey, 1998):

    1. encouragement to explore the environment;
    2. mentoring in basic cognitive and social skills;
    3. celebrating new skills;
    4. rehearsing and expanding new skills;
    5. protection from inappropriate punishment or ridicule for developmental advances; and
    6. stimulation and support in language and symbolic communication.

In fact, the long-lasting impact of those programs has led some researchers to become involved in neuroscience. Clancy Blair of New York University says[1]:

> When I was a graduate student, I worked with Craig Ramey [who was] responsible for the Abecedarian Project. Abecedarian and Perry Preschool and a couple of other projects demonstrated that that provision of that high-quality supervision and support could really make a difference for kids. I thought, "If these programs are having this good long-term effect, they must be affecting some underlying process, and what is that?" That led us to ask a lot of questions about where executive functions come from, how they develop, what are the influences on them and can we use them as a basis for early education.

Emergence of Developmental Cognitive Neuroscience

Research in developmental neuroscience has exploded during the past decades, and it has transformed our understanding of how the brain and behavior develop. In contrast to old-fashioned ideas of maturation as a genetically determined process, new research now shows that development can be seen as a dynamic process of adaptation wherein the brain is constructed in a largely use-dependent fashion (Huttenlocher, 2002; Zelazo, 2013). When we use our brains in particular ways, the neural circuits upon which we rely become more efficient. Fibers connecting regions within a network (and between networks) are myelinated when used, and unused synapses are pruned.

Consistent with early work by Montessori, Vygotsky, Luria, and others, a key focus of research on brain development has been the set of neurocognitive skills supporting agency and autonomy – executive function (EF) skills. Although neuroscience research on reading, math, and social and emotional function also has important implications for education, we focus here on EF skills because these skills play an especially foundational role in learning and because they have been particularly well studied.

EF skills are sets of neurocognitive skills that support the conscious, top-down control of thought, action, and emotion (Diamond, 2013; Zelazo, Blair, & Willoughby, 2016) EF skills depend on increasingly well-understood neural circuits involving brain regions in prefrontal cortex (PFC) and other areas, and they are typically measured behaviorally as three skills: inhibitory control, working memory, and cognitive flexibility (Miyake et al., 2000).

1 Clancy Blair, personal communication with Ellen Galinsky. January 19, 2016.

Inhibitory control involves deliberately suppressing attention (or other responses) to something (e.g., ignoring a distraction or stopping an impulsive utterance). Working memory involves keeping information in mind and, usually, manipulating it in some way. Cognitive flexibility involves thinking about a single stimulus in multiple ways – for example, when considering someone else's perspective on a situation. These skills, which typically work together, are important for characterizing problems, making inferences, and keeping goals and plans in mind despite distractions and interference so that they can be used, deliberately, to guide behavior. As a result, EF skills play a fundamental role in deliberate learning and intentional action, and the atypical development of these skills can lead to widespread and pervasive challenges to healthy adaptation and brain growth (Zelazo, 2015).

Dozens of studies have shown that EF skills are associated with academic achievement (e.g., Allan, Hume, Allan, Farrington, & Lonigan, 2014; Best, Miller, & Naglieri, 2011; Blair & Razza, 2007; McClelland et al., 2014) and socioemotional competence (e.g., Carlson & Moses, 2001; Frye, Zelazo, & Palfai, 1995; Rhoades, Greenberg, & Domitrovich, 2009; Riggs, Jahromi, Razza, Dillworth-Bart, & Müller, 2006). Individual differences in EF skills measured in childhood predict a wide range of outcomes, including school performance and social competence in adolescence (e.g., Mischel, Shoda, & Rodriguez, 1989), college GPA and graduation (Knouse, Feldman, & Blevins, 2014; McClelland, Acock, Piccinin, Rhea, & Stallings, 2013), and physical health, socioeconomic status (SES), and drug-related problems and criminal convictions in adulthood (Moffitt et al., 2011). The predictive power of EF is often greater than that of IQ (e.g., Blair & Razza, 2007), and long-term predictions are seen even when controlling for IQ and childhood SES (e.g., Moffitt et al., 2011).

The human brain is an inherently plastic organ, continually adapting to its environment, but there are periods of relatively high plasticity (often called "sensitive periods") when particular regions of the brain and their corresponding functions are especially susceptible to environmental influences. These periods typically correspond to times of rapid growth in those regions and functions, when relevant neural regions are adapting especially rapidly to structures inherent in the environment (Huttenlocher, 2002). EF skills develop most quickly during early childhood and the transition to adolescence (Zelazo et al., 2013), and these appear to be windows of opportunity for the cultivation of EF-related brain development via well-timed, targeted scaffolding and support (National Academies of Sciences, Engineering, and Medicine, 2019; Zelazo & Carlson, 2012).

As with other neurocognitive skills, the repeated engagement and use of EF skills in problem solving at ever increasing levels of challenge strengthen these skills and increase the efficiency of the corresponding neural circuitry, through processes such as synaptic pruning and myelination (Zelazo & Lee, 2010). A growing body of research has identified specific preschool curricula and activities that improve children's EF skills (Diamond & Ling, 2016). Other research shows that the broader impact of preschool programs is largely mediated through their effect on children's developing EF skills. For example, despite the short-term "fade out" of IQ effects of early childhood interventions (e.g., Puma et al., 2012), longer term follow-ups have revealed that children enrolled in these interventions can show lasting benefits that can be understood in important ways in terms of improved EF skills. For example, these children show higher rates of graduation from high school and better physical and mental health outcomes (e.g., Barnett, 1995). Additionally, Riggs et al. (2006) evaluated the effects of the Promoting Alternative THinking Strategies curriculum (PATHS; Greenberg & Kusché, 1993), a program that explicitly attends to developmental models of prefrontal cortical organization. These authors found that the effects of PATHS on decreasing externalizing and internalizing behaviors were mediated by changes in children's inhibitory control skills.

One limitation of intervention research targeting EF skills concerns generalization or transfer (Melby-Lervåg, Redick, & Hulme, 2016). Although research to date has often failed to find evidence that trained EF skills generalize to different skills or contexts ("far transfer"), most targeted EF interventions have not been optimally designed to promote such far transfer. Research on learning shows clearly that to promote the far transfer of any skill, it is important to train that skill in a wide range of contexts (e.g., Smith, 1982; Schmidt & Bjork, 1992). Newer evidence also indicates the importance not only of practicing EF skills but of encouraging children to reflect upon them metacognitively so that they learn the purpose and utility of EF skills, as well as how and when to apply those skills in new situations (Espinet, Anderson, & Zelazo, 2013; Hadley, Acluche, & Chevalier, 2019; Pozuelos et al., 2018).

Cultivation of EF Skills by Caregivers and Teachers

The development of EF skills during childhood is influenced importantly by caregivers, who support the acquisition of these skills in specific ways. In the context of safe, supportive relationships, children explore challenges. Parents may calibrate these challenges and scaffold children's efforts to meet them. Empirical studies have shown that scaffolding behaviors (i.e., providing children with manageable challenges and offering just enough support for children to succeed) both concurrently and longitudinally predict preschoolers' EF skills (Hammond, Müller, Carpendale, Bibok, & Liebermann-Finestone, 2012; Hughes & Ensor, 2009). Furthermore, parenting behaviors that are autonomy supportive – which include scaffolding and also offering appropriate choices, taking children's perspective and following their lead, and providing encouragement and praise – are also associated with current and future EF skills in young children (Bernier, Carlson, Deschênes, & Matte-Gagné, 2012; Bernier, Carlson, & Whipple, 2010; Distefano, Galinsky, McClelland, Zelazo, & Carlson, 2018; Meuwissen & Carlson, 2015). Within the context of autonomy supportive relationships, children internalize rules and their intrinsic motivation is sustained; this is critical because simply having strong EF skills is not enough, children also need to be motivated to employ them. Finally, parents also provide examples of reflection and goal-directed problem solving.

Teachers are another important scaffolding influence on children's developing EF skills. Across 23 studies, one meta-analysis found a significant correlation between positive teacher–student relationships and children's EF skills (Vandenbroucke, Spilt, Verschueren, Piccinin, & Baeyens, 2018). Specifically, teachers' emotional warmth, teacher–student closeness, and positive behavioral management (e.g., "I like how you are sitting during story time") have been shown to predict gains in preschoolers' EF across the school year (Cadima, Verschueren, Leal, & Guedes, 2016; Fuhs, Farran, & Nesbitt, 2013). Positive teacher–student relationships may be particularly critical for children with lower EF skills who tend to need more structure and support than children with higher EF skills. One study with a sample of 169 Head Start preschoolers demonstrated that teacher emotional support was more strongly associated with gains in EF for children who scored below the 25th percentile on an EF measure in the fall of preschool (Choi et al., 2016). Beyond specific teacher–student relationships, educational environments can be structured to best support EF development by promoting student exploration through engaging and playful trial and error learning, providing opportunities for feedback and reflection, and encouraging children to be accountable for their own learning (e.g., Bonawitz, Shafto, Gweon, Goodman, Spelke, & Schulz, 2011; Bruner, 1966; Marcovitch, Jacques, Boseovski, & Zelazo, 2008; Zimmerman, 1990).

Neuroscience Influenced Interventions in Early Childhood

Basic research on EF development has provided an important foundation for interventions designed to specifically target EF skills in young children, and suggests how to structure places of education to playfully explore their environments in intentional and attentive ways, to practice reflection, and to engage in self-regulated learning. These neuroscience-based interventions span multiple contexts, from the home to the classroom, and, increasingly, aim to connect these developmental systems to have the greatest impact on children.

Lab-based Interventions

In a laboratory context, a brief psychological distancing intervention has been shown to effectively boost EF performance in the moment. Psychological distancing refers to putting mental space between oneself and the current moment. White and Carlson (2016) found that when 5-year-olds were asked to pretend to be their favorite media character (e.g., Batman), thus providing some psychological distance, they performed better on an EF task compared with children who were asked to think about themselves during the task, as well as children in a no-manipulation control condition. This study reveals the importance of pretending to the development of EF skills. In addition, psychological distancing is theorized to be effective because it promotes reflection, and thereby supports children's intentional use of EF skills. In fact, one study with preschoolers (Espinet et al., 2013) studied the effect of a brief 15-minute reflection training session, in which children who failed an EF task (the Dimensional Change Card Sort, or DCCS; Zelazo, 2006) were given a different version of the DCCS (with different shapes and colors) and taught to pause before responding, reflect on the hierarchical nature of the task, and formulate higher-order rules for responding flexibly. Compared with children who received only *minimal yes/no feedback* (without practice in reflection) and with children who received *mere practice with no feedback* at all, children who received *reflection training* showed significant improvements in performance on a subsequent administration of the DCCS. The benefits of reflection training generalized to improvements on a measure of flexible perspective taking (a false belief task) – an example of far transfer – and these behavioral changes were accompanied by changes in children's brain activity, as measured using electroencephalography (EEG). Moriguchi, Sakata, Ishibashi, and Ishikawa (2015) also encouraged preschoolers to reflect on the DCCS, by asking them to teach a puppet the rules of the task. Compared with controls, trained children showed improvement in performance on the DCCS along with increased brain activity in parts of prefrontal cortex.

Interventions in the Classroom

Tools of the Mind is a research-based early childhood model combining teacher professional development with a comprehensive curriculum that helps PreK- and Kindergarten-aged children develop cognitive, social-emotional, self-regulatory, and foundational academic skills. Their approach includes the use of specific tactics to support memory and learning, as well as the organization of "shared cooperative activity" designed to promote social-emotional as well as cognitive development. Tools of the Mind focuses on EF skills as a primary mechanism through which children make academic progress and develop social competencies. In order to promote these skills, the program blends teacher-led scaffolding of a comprehensive curriculum of literacy, mathematics, and science activities aligned with the Common Core Standards as well as with child-directed activities and, importantly, structured sociodramatic play. In this curriculum, teachers help children build self-regulation skills through purposeful playful and engaging interactions with classmates.

A study by Blair and Raver (2014) found that the Tools of the Mind Kindergarten program had positive effects on EF skills, reasoning ability, the control of attention, and levels of salivary cortisol and alpha amylase (i.e., indicators of neuroendocrine function). Results also demonstrated improvements in reading, vocabulary, and mathematics at the end of kindergarten that increased into the first grade. As the researchers state, "A number of effects were specific to high-poverty schools, suggesting that a focus on executive functions and associated aspects of self-regulation in early elementary education holds promise for closing the achievement gap."

Other EF interventions in a school setting have been more child-focused and incorporated classroom activities specifically designed to provide children with opportunities to practice their EF skills. In the semi-structured block play intervention described in Schmitt, Korucu, Napoli, Bryant, and Purpura (2018), researchers worked in small groups with preschool-aged children twice per week for 7 weeks. The block play intervention required children to plan and to solve problems while working collaboratively with peers. Results indicated a moderate effect size of the intervention on children's EF skills, with the intervention being most effective for children from low socioeconomic backgrounds.

A similar child-focused intervention leveraged the circle time format typical in most preschool classrooms. Two times per week for 8 weeks, children were led by a trained assistant through a series of games designed to exercise children's EF skills (Tominey & McClelland, 2011). For example, in the freeze dance game, children danced while music was playing, but then stopped when the music was paused; this activity requires children to use their inhibitory control skills because they need to stop and start their behaviors based on the rules of the game. One study evaluated the effect of the Circle Time Games intervention in a sample of 276 children enrolled in Head Start classrooms (Schmitt, McClelland, Tominey, & Acock, 2015). Results showed significant effects of the intervention on the two measures of EF used in the study. Furthermore, the intervention appeared to be most effective for children who were English language learners. Together, these two child-focused interventions suggest that one way to improve children's EF skills is to give them targeted, supported, and engaging opportunities to practice these skills.

Parenting Interventions

The Attachment and Biobehavioral Catchup (ABC) intervention is designed to strengthen attachment relationships and support effective co-regulation strategies in foster parents and children (Lewis-Morrarty, Dozier, Bernard, Terracciano, & Moore, 2012). Compared with children who received an active control condition, children who received the ABC intervention showed significant improvements in their EF skills. Another intervention conducted by Obradović and colleagues (Obradović, Yousafzai, Finch, & Rasheed, 2016) targeted sensitive and responsive parenting through community groups and home visiting in mothers and their infants living in rural Pakistan. These authors found that the intervention had both a direct effect on children's EF skills at age 4 years and an indirect effect through maternal scaffolding behaviors, suggesting that parenting behaviors are an important mediator of intervention effects.

Interventions Using Technology

Vroom, a program of the Bezos Family Foundation, helps parents transform everyday routines, such as bed time, bath time, meal time, into brain-building moments by providing a validating message – *you have what it takes* – through science-backed tips that parents can use in the time they have with their children. There are more than 1,000 tips for families for children birth through 5, which are available online through a mobile phone-based App, and by text at no cost to parents. These tips have been designed to be accessible to parents and families from diverse backgrounds: They are available in 14 languages, written at a fifth-grade reading level or lower, and contain no more than 300 characters (Galinsky, Bezos, McClelland, Carlson, & Zelazo, 2017). Many of the tips allow children to directly practice their developing EF skills. For example, in one daily tip, parents are encouraged to ask their children to reflect on their day, but to try to do so backwards. This activity requires EF because children need to hold in mind what they did during the day, but manipulate the information to go in reverse order. They also need to inhibit the usual way they tell their parents about their day. In addition to the tip itself, Vroom provides information on how the activity relates to children's cognitive development as a way to further connect families to developmental science.

Vroom currently reaches more than one million families via local community activations in over 170 sites with a total reach of 37 states. It also reaches families through partnerships with brands, with spots on Daniel Tiger and a mini-series on Univision. Vroom has 860,000 digital followers. There are seven international sites, including work through the International Rescue Committee for Syrian refugee families. Although the impact of Vroom on children's developing EF skills has yet to be assessed, this approach has considerable promise and is designed to be scalable.

Two-generation Interventions

One intervention framework that has been used to integrate many of the effective EF intervention strategies is the two-generation approach. First gaining attention decades ago, two-generation interventions target both children and their families to bolster positive developmental outcomes (Chase-Lansdale & Brooks-Gunn, 2014; Ramey, Ramey, Gaines, & Blair, 1995; Shonkoff & Fisher, 2013). Theory in support of this approach posits that development occurs within the context of multiple interacting systems (Bronfenbrenner, 1992; Ford & Lerner, 1992). Thus, interventions targeting children's development cannot ignore the influences of these developmental systems. Even the most well-designed child-focused interventions may come undone if children return to a home environment with problematic parent–child relationships and chaotic conditions. Despite the strong theoretical rationale for two-generation approaches, however, there are only a few interventions that have focused on parents and children together.

The Parents and Children Making Connections – Highlighting Attention (PCMC-A) program was designed for parents and preschool-aged children to improve parent and child self-regulation (Neville et al., 2013). Parents attend weekly 2-hour group sessions for 8 weeks while preschool-aged children participate in separate small groups to practice attentional and emotional regulation. For example, to practice attentional regulation, children are told to color in the lines of a complex picture while other children attempt to distract them (e.g., tossing balloons in the air). Results indicated that children showed changes in the neural systems underlying selective attention and parents reported less stress and demonstrated more positive parenting behaviors in an observational parenting task compared with both a child-focused active control group (i.e., Attention Boost for Children) and a Head Start-only control group. These findings suggest that a two-generation approach is promising for strengthening both parenting outcomes and children's brain development. Additional research is needed, however, to create and test two-generation interventions that show effects on behavioral measures of children's EF skills and engage multiple developmental systems for more ubiquitous intervention effects.

In an attempt to mitigate some of the neurobiological effects of early adversity, Fisher, Frenkel, Noll, Berry, and Yockelson (2016) developed the Filming Interactions to Nurture Development (FIND) program. FIND is a two-generation intervention that targets the early caregiving relationship because positive infant–caregiver experiences are central to healthy brain development during this sensitive period of development. The FIND program uses video feedback during parent–child interactions to promote five elements of serve and return: sharing focus, supporting and encouraging, naming, back and forth interaction, and beginnings and endings. These intervention targets are hypothesized to increase parents' own EF skills (e.g., parents practice their inhibitory control skills when they wait for their child's cues), which in turn support future positive parenting and mental well-being, as well as children's neurocognitive and socioemotional skills. Trials of the FIND intervention are currently being implemented in childcare settings and home-visiting programs.

A Civic Science Approach to Interventions

Mind in the Making at the Bezos Family Foundation is an ongoing effort to curate the science of children's brain development and learning, to share it with the general public, families, and professionals, to translate this research into action (training and tools), and to evaluate its impact. Among its key initiatives are its Mind in the Making Learning Modules – a 16-hour training program created through a civic science approach. A civic science framework recognizes scientific research as a pragmatic human activity designed to support evidence-based solutions to societal needs. This framework views parents, teachers, and other early childhood professionals as essential partners in the process of identifying and responding appropriately to the needs of children, and it starts by fostering connections that may not exist. For example, parents from lower socioeconomic backgrounds may report strained relationships with professionals, such as teachers or researchers (e.g., Waanders, Mendez, & Downer, 2007). The Mind in the Making group trainings bring together parents and professionals across sectors serving children birth through 8 years (early education, schools, juvenile justice, parent support and education, museums and libraries, etc.) to learn about EF, autonomy support, and essential life skills that are supported by EF skills in both caregivers and their children. In this way, the Modules build relationships among the people across sectors in order to break down silos. The Modules also redefine parent engagement by inviting families and professionals to learn together, thus moving away from the notion of professionals "teaching" parents.

The seven life skills are focus and self-control, perspective-taking, communicating, making connections, critical thinking, taking on challenges, and self-directed engaged learning. The Modules make connections among these seven essential skills and underlying executive functions. For example, parents and professionals learn that perspective-taking requires EF skills because one must inhibit one's own thoughts and feelings and instead view a situation from another person's point of view. They also learn that perspective-taking can be improved in themselves by pausing to think about the other person before acting, as well as in their children by talking with them about what others might be thinking about. The Modules promote "facilitated learning," eschewing the notion of training (top-down learning) in favor of learning from everyone involved – community leaders and scientists – in a respectful, experiential way. Throughout the Modules are videos of leading developmental and neuroscientists that are virtual field trips to their labs, where the scientists share their questions and how they pursue answers. Unique to the Modules is that they begin with adults, promoting their own life skills before promoting these skills in children so that the participants can see and feel the value of these skills in their own lives. The Modules include other materials developed by Mind in the Making, including the Vroom tips, Skill Building Opportunities – a tool that address parents' and professionals' frequently asked questions with strategies that help them move from managing children's challenging behavior to promoting life skills – and case studies where participants practice autonomy-supportive caregiving. Additional themes of the sessions include promoting goal setting, based on the research of Gabriele Oettingen (2015). The overall aim of this program is to create a "community of learners" – ongoing learners about oneself and others in order to keep the fire for learning – the inherent curiosity and need for growth that Montessori recognized – alive.

Mind in the Making Learning Modules are available in 20 states and through the National Head Start Association. An evaluation in Providence, Rhode Island, where MITM Modules are used in all 22 elementary schools, found that parents expressed more confidence in their own ability to promote the academic skills of their children and in the importance of their involvement in the education of their children. They also expressed less authoritarian views. Educator surveys have indicated significant shifts in their attitudes toward parent involvement, parental inclusion, and developmentally appropriate beliefs regarding young children.

Conclusions

Although neuroscience is a relatively new player in early education, it has transformed our understanding of the conditions that support learning and brain development. By enabling scientists and educators alike to understand learning from both a brain-based and a behavioral perspective, neuroscience has drawn attention to issues such as neuroplasticity, sensitive periods in development, the importance of EF skills for learning and adaptation, and how to cultivate not only the development of these skills but also their transfer to new situations. In addition, by providing greater precision in the mechanisms underlying effective interventions, neuroscience can accelerate the effective refinement and perhaps personalization of these interventions.

As with any new influence, there are some critical concerns (Horn, Fisher, Pfeifer, Allen, & Berkman, 2020). One has only to attend to educational conferences today to see that numerous products and programs are advertised as "brain building," as taking a "train the brain approach," or as "brain-based education," whether or not they were designed and tested using rigorous neuroscientific tools. Additionally, there are concerns that EF skills will be taught in a drill and practice way – not an engaging and playful way – and that children will be judged on their EF skills (much in the way that they were judged on presumed IQ in the past).

As neuroscience become more mainstream, it is essential to guard against those making false claims or using narrow practices so that the promise of neuroscience can add to – rather than detract from – helping children and families thrive and the profession of early education flourish.

References

Allan, N.P., Hume, L.E., Allan, D.M., Farrington, A.L., & Lonigan, C.J. (2014). Relations between inhibitory control and the development of academic skills in preschool and kindergarten: A meta-analysis. *Developmental Psychology*, 50, 2368–2379. doi:10.1037/a0037493

Barnett, W.S. (1995). Long-term effects of early childhood programs on cognitive and school outcomes. *The Future of Children*, 5, 25–50. doi:10.2307/1602366

Bernier, A., Carlson, S.M., Deschênes, M., & Matte-Gagné, C. (2012). Social factors in the development of early executive functioning: A closer look at the caregiving environment. *Developmental Science*, 15, 12–24. doi:10.1111/j.1467-7687.2011.01093.x

Bernier, A., Carlson, S.M., & Whipple, N. (2010). From external regulation to self-regulation: Early parenting precursors of young children's executive functioning. *Child Development*, 8, 326–339. doi:10.1111/j.1467-8624.2009.01397.x

Best, J.R., Miller, P.H., & Naglieri, J.A. (2011). Relations between executive function and academic achievement from ages 5 to 17 in a large, representative national sample. *Learning and Individual Differences*, 21, 327–336. doi:10.1016/j.lindif.2011.01.007

Blair, C., & Raver, C.C. (2014). Closing the achievement gap through modification of neurocognitive and neuroendocrine function: Results from a cluster randomized controlled trial of an innovative approach to the education of children in kindergarten. *PLoS ONE*, 9, e112393. doi:10.1371/journal.pone.0112393

Blair, C., & Razza, R.P. (2007). Relating effortful control, executive function, and false belief understanding to emerging math and literacy ability in kindergarten. *Child Development*, 78, 647–663. doi:10.1111/j.1467-8624.2007.01019.x

Bonawitz, E., Shafto, P., Gweon, H., Goodman, N.D., Spelke, E., & Schulz, L. (2011). The double-edged sword of pedagogy: Instruction limits spontaneous exploration and discovery. *Cognition*, 120, 322–330. doi:10.1016/j.cognition.2010.10.001

Bronfenbrenner, U. (1992). Ecological systems theory. In R. Vasta (Ed.), *Six theories of child development: Revised formulations and current issues* (pp. 187–249). London: Jessica Kingsley Publishers.

Bruner, J.S. (1966). *Toward a theory of instruction*. Cambridge, MA: Harvard University Press.

Cadima, J., Verschueren, K., Leal, T., & Guedes, C. (2016). Classroom interactions, dyadic teacher–child relationships, and self-regulation in socially disadvantaged young children. *Journal of Abnormal Child Psychology*, 44, 7–17. doi:10.1007/s10802-015-0060-5

Carlson, S.M., & Moses, L.J. (2001). Individual differences in inhibitory control and children's theory of mind. *Child Development*, 72, 1032–1053. doi:10.1111/1467-8624.00333

Carnegie Task Force. (1994). *Starting points: Meeting the needs of our youngest children*. New York, NY: Author.

Chase-Lansdale, P.L., & Brooks-Gunn, J. (2014). Two-generation programs in the twenty-first century. *Future of Children*, 24, 13–39. doi:10.1353/foc.2014.0003

Choi, J.Y., Castle, S., Williamson, A.C., Young, E., Worley, L., Long, M., & Horm, D.M. (2016). Teacher–child interactions and the development of executive function in preschool-age children attending Head Start. *Early Education and Development*, 27, 751–769. doi:10.1080/10409289.2016.1129864

Diamond, A. (2013). Executive functions. *Annual Review of Psychology*, 64, 135–168. doi:10.1146/annurev-psych-113011-143750

Diamond, A., & Ling, D.S. (2016). Conclusions about interventions, programs, and approaches for improving executive functions that appear justified and those that, despite much hype, do not. *Developmental Cognitive Neuroscience*, 18, 34–48. doi:10.1016/j.dcn.2015.11.005

Distefano, R., Galinsky, E., McClelland, M.M., Zelazo, P.D., & Carlson, S.M. (2018). Autonomy-supportive parenting and associations with child and parent executive function. *Journal of Applied Developmental Psychology*, 58, 77–85. doi:10.1016/j.appdev.2018.04.007

Espinet, S.D., Anderson, J.E., & Zelazo, P.D. (2013). Reflection training improves executive function in preschool-age children: Behavioral and neural effects. *Developmental Cognitive Neuroscience*, 4, 3–15. doi:10.1016/j.dcn.2012.11.009

Fisher, P.A., Frenkel, T.I., Noll, L.K., Berry, M., & Yockelson, M. (2016). Promoting healthy child development via a two-generation translational neuroscience framework: The Filming Interactions to Nurture Development video coaching program. *Child Development Perspectives*, 10, 251–256. doi:10.1111/cdep.12195

Ford, D.H., & Lerner, R.M. (1992). *Developmental systems theory: An integrative approach*. Thousand Oaks, CA: Sage.

Frye, D., Zelazo, P.D., & Palfai, T. (1995). Theory of mind and rule-based reasoning. *Cognitive Development*, 10, 483–527. doi:10.1016/0885-2014(95)90024-1

Fuhs, M.W., Farran, D.C., & Nesbitt, K.T. (2013). Preschool classroom processes as predictors of children's cognitive self-regulation skills development. *School Psychology Quarterly*, 28, 347–359. doi:10.1037/spq0000031

Galinsky, E. (2006). *The economic benefits of high-quality early childhood programs: What makes the difference?* Washington, DC: Committee for Economic Development.

Galinsky, E., Bezos, J., McClelland, M., Carlson, S.M., & Zelazo, P.D. (2017). Civic science for public use: Mind in the Making and Vroom. *Child Development*, 88, 1409–1418. doi:10.1111/cdev.12892

Greenberg, M.T., & Kusché, C.A. (1993). *Promoting social and emotional development in deaf children: The PATHS project*. Seattle, WA: University of Washington Press.

Hadley, L.V., Acluche, F., & Chevalier, N. (2019). Encouraging performance monitoring promotes proactive control in children. *Developmental Science*, e12861. doi:10.1111/desc.12861

Hammond, S.I., Müller, U., Carpendale, J.I.M., Bibok, M.B., & Liebermann-Finestone, D.P. (2012). *The effects of parental scaffolding on preschoolers' executive function*. *Developmental Psychology*, 48, 271–281. doi:10.1037/a0025519

Heckman, J.J., Moon, S.H., Pinto, R., Savelyev, P.A., & Yavitz, A. (2010). The rate of return to the HighScope Perry Preschool Program. *Journal of Public Economics*, 94, 114–128. doi:10.1016/j.jpubeco.2009.11.001

Horn, S.R., Fisher, P.A., Pfeifer, J.H., Allen, N.B., & Berkman, E.T. (2020). Levers and barriers to success in the use of translational neuroscience for the prevention and treatment of mental health and promotion of well-being across the lifespan. *Journal of Abnormal Psychology*, 129(1), 38–48. doi:10.1037/abn0000465

Hughes, C., & Ensor, R. (2009). How do families help or hinder the emergence of early executive function? *New Directions for Child and Adolescent Development*, 123, 35–50. doi:10.1002/cd

Huttenlocher, P.R. (2002). *Neural plasticity: The effects of environment on the development of the cerebral cortex*. Cambridge, MA: Harvard University Press.

Knouse, L.E., Feldman, G., & Blevins, E.J. (2014). Executive functioning difficulties as predictors of academic performance: Examining the role of grade goals. *Learning and Individual Differences*, 36, 19–26. doi:10.1016/j.lindif.2014.07.001

Lewis-Morrarty, E., Dozier, M., Bernard, K., Terracciano, S.M., & Moore, S.V. (2012). Cognitive flexibility and theory of mind outcomes among foster children: Preschool follow-up results of a randomized clinical trial. *Journal of Adolescent Health*, 51, S17–S22. doi:10.1016/j.jadohealth.2012.05.005

Lillard, A., & Else-Quest, N. (2006). The early years: Evaluating Montessori education. *Science*, 313, 1893–1894. doi:10.1126/science.1132362

Luria, A.R. (1961). *The role of speech in the regulation of normal and abnormal behaviour*. New York, NY: Pergamon Press.

Marcovitch, S., Jacques, S., Boseovski, J.J., & Zelazo, P.D. (2008). Self-reflection and the cognitive control of behavior: Implications for learning. *Mind, Brain, and Education*, 2, 136–141.

McClelland, M.M., Acock, A.C., Piccinin, A., Rhea, S.A., & Stallings, M.C. (2013). Relations between preschool attention span-persistence and age 25 educational outcomes. *Early Childhood Research Quarterly*, 28, 314–324. doi:10.1016/j.ecresq.2012.07.008

McClelland, M.M., Cameron, C.E., Duncan, R., Bowles, R.P., Acock, A.C., Miao, A., & Pratt, M.E. (2014). Predictors of early growth in academic achievement: The head-toes-knees-shoulders task. *Frontiers in Psychology*, 5, 1–14. doi:10.3389/fpsyg.2014.00599

Melby-Lervåg, M., Redick, T.S., & Hulme, C. (2016). Working memory training does not improve performance on measures of intelligence or other measures of "far transfer": Evidence from a meta-analytic review. *Perspectives on Psychological Science*, 11, 512–534. doi:10.1177/1745691616635612

Meuwissen, A.S., & Carlson, S.M. (2015). Fathers matter: The role of father parenting in preschoolers' executive function development. *Journal of Experimental Child Psychology*, 140, 1–15. doi:10.1016/j.jecp.2015.06.010

Mischel, W., Shoda, Y., & Rodriguez, M.L. (1989). Delay of gratification in children. *Science*, 244, 933–938. doi:10.1126/science.2658056

Miyake, A., Friedman, N.P., Emerson, M.J., Witzki, A.H., Howerter, A., & Wager, T.D. (2000). The unity and diversity of executive functions and their contributions to complex "frontal lobe" tasks: A latent variable analysis. *Cognitive Psychology*, 41, 49–100. doi:10.1006/cogp.1999.0734

Moffitt, T.E., Arseneault, L., Belsky, D., Dickson, N., Hancox, R.J., & Harrington, H. (2011). A gradient of childhood self-control predicts health, wealth, and public safety. *Proceedings of the National Academy of Sciences*, 108, 2693–2698. doi:10.1073/pnas.1010076108

Montessori, M. (1912). *The Montessori method: scientific pedagogy as applied to child education in the children's houses*. New York, NY: Frederick A. Stokes.

Moriguchi, Y., Sakata, Y., Ishibashi, M., & Ishikawa, Y. (2015). Teaching others rule use improves executive function and prefrontal activations in young children. *Frontiers in Psychology*, 6, 894. doi:10.3389/fpsyg.2015.00894

National Academies of Sciences, Engineering, and Medicine. (2019). The promise of adolescence: Realizing opportunity for all youth. Washington, DC: The National Academies Press. doi:10.17226/25388

Neville, H.J., Stevens, C., Pakulak, E., Bell, T.A., Fanning, J., Klein, S., & Isbell, E. (2013). Family-based training program improves brain function, cognition, and behavior in lower socioeconomic status preschoolers. *Proceedings of the National Academy of Sciences*, 110, 12138–12143. doi:10.1073/pnas.1304437110

Obradović, J., Yousafzai, A.K., Finch, J.E., & Rasheed, M.A. (2016). Maternal scaffolding and home stimulation: Key mediators of early intervention effects on children's cognitive development. *Developmental Psychology*, 52, 1409–1421. doi:10.1037/dev0000182

Oettingen, G. (2015). *Rethinking positive thinking: Inside the new science of motivation*. New York, NY: Penguin Random House.

Pozuelos, J.P., Combita, L.M., Abundis, A., Paz-Alonso, P.M., Conejero, Á., Guerra, S., & Rueda, M. R. (2019). Metacognitive scaffolding boosts cognitive and neural benefits following executive attention training in children. *Developmental Science*, 22, e12756. doi:10.1111/desc.12756

Puma, M., Bell, S., Cook, R., Heid, C., Broene, P., Jenkins, F., Mashburn, A., & Downer, J. (2012). *Third Grade Follow-up to the Head Start impact study final report* (OPRE Report # 2012–2045). Washington, DC: Office of Planning, Research and Evaluation, Administration for Children and Families, U.S. Department of Health and Human Services. Retrieved from https://www.acf.hhs.gov/opre/resource/third-grade-follow-up-to-the-head-start-impact-study-final-report

Ramey, C.T. (2018). The Abecedarian approach to social, educational, and health disparities. *Clinical Child and Family Psychology Review*, 21, 527–544. doi:10.1007/s10567-018-0260-y

Ramey, C.T., & Ramey, S.L. (1998). Early intervention and early experience. *American Psychologist*, 53, 109–120. doi:10.1037/0003-066X.53.2.109

Ramey, C., Ramey, S., Gaines, K., & Blair, C. (1995). Two-generation early intervention programs: A child development perspective. In S. Smith (Ed.), *Two-generation programs for families in poverty: A new intervention strategy* (pp. 199–228). Norwood, NJ: Ablex.

Rhoades, B.L., Greenberg, M.T., & Domitrovich, C.E. (2009). The contribution of inhibitory control to preschoolers' social-emotional competence. *Journal of Applied Developmental Psychology*, 30, 310–320. doi:10.1016/j.appdev.2008.12.012

Riggs, N.R., Jahromi, L.B., Razza, R.P., Dillworth-Bart, J.E., & Mueller, U. (2006). Executive function and the promotion of social-emotional competence. *Journal of Applied Developmental Psychology*, 27, 300–309. doi:10.1016/j.appdev.2006.04.002

Schmidt, R.A., & Bjork, R.A. (1992). New conceptualizations of practice: Common principles in three paradigms suggest new concepts for training. *Psychological Science*, 3, 207–218. doi:10.1111/j.1467-9280.1992.tb00029.x

Schmitt, S.A., Korucu, I., Napoli, A.R., Bryant, L.M., & Purpura, D.J. (2018). Using block play to enhance preschool children's mathematics and executive functioning: A randomized controlled trial. *Early Childhood Research Quarterly*, 44, 181–191. doi:10.1016/j.ecresq.2018.04.006

Schmitt, S.A., McClelland, M.M., Tominey, S.L., & Acock, A.C. (2015). Strengthening school readiness for Head Start children: Evaluation of a self-regulation intervention. *Early Childhood Research Quarterly*, 30, 20–31. doi:10.1016/j.ecresq.2014.08.001

Shonkoff, J.P., & Fisher, P.A. (2013). Rethinking evidence-based practice and two-generation programs to create the future of early childhood policy. *Development and Psychopathology*, 25, 1635–1653. doi:10.1017/S0954579413000813

Shore, R. (1997). *Rethinking the brain: New insights into early development.* New York, NY: Families and Work Institute.

Smith, S.M. (1982). Enhancement of recall using multiple environmental contexts during learning. *Memory & Cognition*, 10, 405–412. doi:10.3758/BF03197642

Tominey, S.L., & McClelland, M.M. (2011). Red light, purple light: Findings from a randomized trial using circle time games to improve behavioral self-regulation in preschool. *Early Education and Development*, 22, 489–519. doi:10.1080/10409289.2011.574258

Vandenbroucke, L., Spilt, J., Verschueren, K., Piccinin, C., & Baeyens, D. (2018). The classroom as a developmental context for cognitive development: A meta-analysis on the importance of teacher–student interactions for children's executive functions. *Review of Educational Research*, 88, 125–164. doi:10.3102/0034654317743200

Vygotsky, L.S. (1929). II. The problem of the cultural development of the child. *The Pedagogical Seminary and Journal of Genetic Psychology*, 36, 415–434. doi:10.1080/08856559.1929.10532201

Vygotsky, L.S. (1934/1962). *Thought and language.* Cambridge, MA: MIT Press.

Waanders, C., Mendez, J.L., & Downer, J.T. (2007). Parent characteristics, economic stress and neighborhood context as predictors of parent involvement in preschool children's education. *Journal of School Psychology*, 45, 619–636. doi:10.1016/j.jsp.2007.07.003

White, R.E., & Carlson, S.M. (2016). What would Batman do? Self-distancing improves executive function in young children. *Developmental Science*, 19, 419–426. doi:10.1111/desc.12314

Zelazo, P.D. (2006). The dimensional change card sort (DCCS): A method of assessing executive function in children. *Nature Protocols*, 1, 297–301. doi:10.1038/nprot.2006.46

Zelazo, P.D. (2013). Developmental psychology: A new synthesis. In P.D. Zelazo (Ed.), *Oxford handbook of developmental psychology (Volume 1: Body and mind; Volume 2: Self and other)* (pp. 3–12). New York, NY: Oxford University Press.

Zelazo, P.D. (2015). Executive function: Reflection, iterative reprocessing, complexity, and the developing brain. *Developmental Review*, 38, 55–68. doi:10.1016/j.dr.2015.07.001

Zelazo, P.D., Anderson, J.E., Richler, J., Wallner-Allen, K., Beaumont, J.L., & Weintraub, S. (2013). NIH Toolbox Cognition Battery (CB): Measuring executive function and attention. *Monographs of the Society for Research in Child Development*, 78(4), 16–33. doi:10.1111/mono.12032

Zelazo, P.D., Blair, C.B., & Willoughby, M.T. (2016). Executive function: Implications for education (NCER 2017–2000). Washington, DC: National Center for Education Research, Institute of Education Sciences, U.S. Department of Education. Retrieved from https://ies.ed.gov/ncer/pubs/20172000/pdf/20172000.pdf

Zelazo, P.D., & Carlson, S.M. (2012). Hot and cool executive function in childhood and adolescence: Development and plasticity. *Child Development Perspectives*, 6, 354–360. doi:10.1111/j.1750-8606.2012.00246.x

Zelazo, P.D., & Lee, W.S.C. (2010). Brain development: An overview. In W.F. Overton (Ed.), *Cognition, biology, and methods across the lifespan. Volume 1 of the Handbook of life-span development* (pp. 89–114). Hoboken, NJ: Wiley.

Zimmerman, B.J. (1990). Self-regulated learning and academic achievement: An overview. *Educational Psychologist*, 25, 3–17. doi:10.1207/s15326985ep2501_2

14

CONCLUDING THOUGHTS

M. Elizabeth Graue

Thus, the scientific truths of the twentieth century seem to have a much shorter life-span than those of the last century because scientific activity is now much greater. If, in the next century, scientific activity increases tenfold, then the life expectancy of any scientific truth can be expected to drop to perhaps one-tenth as long as now. What shortens the life-span of the existing truth is the volume of hypotheses offered to replace it; the more the hypotheses, the shorter the time span of the truth. And what seems to be causing the number of hypotheses to grow in recent decades seems to be nothing other than scientific method itself. The more you look, the more you see (Pirsig, 1974, p. 101).[1]

Science's embrace of early childhood education has intensified and broadened in the twentieth and twenty-first century, from the Child Study movement that created norms in child development to the prodigious research analyses of data from the Early Childhood Longitudinal Study. Though some have argued that early education exists in the shadow of the science of child development (Bloch, 1992), as the chapters in this volume demonstrate, the topic of early childhood education has captured the attention of researchers from a variety of disciplines. Their tools and disciplinarily informed perspectives have created new knowledge of different facets of the field, they have pulled early education on the national stage, and they have multiplied the sheer number of journal articles with early childhood education in the title. It is a good time to be in early childhood education.

But following Pirsig, is all this scientific attention to early childhood building a field or is it speeding up knowledge production and fragmenting it? What are the implications of looking more and seeing more? This book is an attempt to answer these questions. By bringing together scholars from many disciplines to consider how their traditions have framed the problem of early childhood education, we have the opportunity to see an inverted version of the blind men and the elephant story. Rather than attempting to figure out the early childhood elephant, authors have used their disciplinary perspectives to define early childhood education. This is important to recognize because *how* we look shapes *what* we see. And the authors help us see that early childhood education is many things. The chapters reflect an early childhood that is variously social, individual, interactional, cultural, historical, economic, biological, and neurological, making what we might have thought of as a unitary and quite

1 Some might question using the fictional work of a 1970s' philosopher who wrote a mystical novel as a rhetorical tool in an academic book. But I thank Robert Stake for introducing me to Pirsig in my first evaluation course. Sometimes fiction can illuminate things in ways that science cannot.

common entity something much more complicated. How the authors define the problem space of early childhood education creates a version of the field that is much a reflection of a discipline as it is of the thing itself.

This would not be a challenge if our common view of science did not suggest that it works toward convergence rather than divergence. I hear this view when my colleagues say, "We know that... ." citing aggregated facts that reflects a deep belief in the notions of generalizability and universality. From this perspective, we should be building toward a set of knowns that will guide future research, policy, and practice. This stance, reflected in many chapters here, has been part of the engine that has driven the substantial investment in research on early childhood education and the subsequent investment in programs. All of this has worked in our favor and has created many opportunities for the field. It is likely that some purchase this book hoping it will provide a scientific narrative to explain early childhood.

But what if seeing more leads to knowing more that is *contingent* and even *temporary*, as Pirsig suggests? This line of argument might seem antithetical for a co-editor of a book that is exploring how the sciences have influenced early childhood education. But I would argue that it is possible and perhaps even necessary to embrace a view that reflects the multiplicity and contingency of science. In a paper that calls for more diverse research and researchers, Edmund Gordon (1985) argued that the "search for universals has inhibited rather than enhanced the encirclement of social science knowledge" (p. 117). He continued:

> Explication of human behavior is dependent on the investigator's interpretation of the origins of the behavior, the values placed on the behavior and the behaving persons, as well as on the interpretations of the behavior itself. Thus, theoretical proposals and laws are not value-free. The questions and problems of interest to the investigators generally reflect the theoretical bias of the investigator.
>
> *(p. 121)*

Recognizing the basic power of values that shapes sciences does not need to be framed as a problem and "radically different perspectives on scientific questions need not be dismissed solely on the ground of their uniqueness" (p. 123). These differences can open new conversations, create sites of growth, and promote new research directions. If we are lucky, they will make positive differences in children's lives. But that is only possible if the field of early childhood education and those who bring their expertise to bear on the subject recognize science's limitations for creating a map for policy and practice and that it not only draws people together, but also keeps people out.

Gathering a variety of perspectives on early childhood education is a first step in broadening the conversation. But it is only a first step. To end this chapter, I go back to the beginning of the book and Barbara Bowman's Foreword. She reinforces Gordon's attention to values in research and the need for more viewpoints and voices if the field is to move forward. Though shining the bright light of science on early childhood education has allowed us to see more:

> It is essential for professionals to develop a sound knowledge base that can be viewed against a backdrop of social values. Science is not value free. Scientific research is not value free. We will need to be intentional and vigilant to ensure that new research reflects the field's commitment to diversity and the connections between values, science, and research is transparent.

References

Bloch, M. (1992). Critical perspectives on the historical relationship between child development and early childhood education research. In S. Kessler & B.B. Swadener (Eds.), *Reconceptualizing early childhood curriculum*. New York: Teachers College Press.

Gordon, E.W. (1985). Social science knowledge production and minority experiences. *The Journal of Negro Education*, 54(2), 117. doi:10.2307/2294927

Pirsig, R.M. (1974). *Zen and the art of motorcycle maintenance. An inquiry into values*. Toronto: Bantam Books.